Zurich Lectures in Advanced Mathematics

Edited by

Mathematics in Zurich has a long and distinguished tradition, in which the writing of lecture notes volumes and research monographs play a prominent part. The *Zurich Lectures in Advanced Mathematics* series aims to make some of these publications better known to a wider audience. The series has three main constituents: lecture notes on advanced topics given by internationally renowned experts, graduate text books designed for the joint graduate program in Mathematics of the ETH and the University of Zurich, as well as contributions from researchers in residence at the mathematics research institute, FIM-ETH. Moderately priced, concise and lively in style, the volumes of this series will appeal to researchers and students alike, who seek an informed introduction to important areas of current research.

Previously published in this series:

Yakov B. Pesin, *Lectures on partial hyperbolicity and stable ergodicity*
Sun-Yung Alice Chang, *Non-linear Elliptic Equations in Conformal Geometry*
Sergei B. Kuksin, *Randomly forced nonlinear PDEs and statistical hydrodynamics in 2 space dimensions*
Pavel Etingof, *Calogero-Moser systems and representation theory*
Guus Balkema and Paul Embrechts, *High Risk Scenarios and Extremes - A geometric approach*

Published with the support of the Huber-Kudlich-Stiftung, Zürich

Demetrios Christodoulou

Mathematical Problems of General Relativity I

European Mathematical Society

Author:

Prof. Demetrios Christodoulou
Department of Mathematics
ETH-Zentrum
8092 Zürich
Switzerland

2000 Mathematics Subject Classification (primary; secondary): 83C05; 35L70, 35Q75, 53C50, 58J45, 83C40, 83C45

ISBN 978-3-03719-005-0

The Swiss National Library lists this publication in The Swiss Book, the Swiss national bibliography, and the detailed bibliographic data are available on the Internet at http://www.helveticat.ch.

Contact address:

European Mathematical Society Publishing House
Seminar for Applied Mathematics
ETH-Zentrum FLI C4
CH-8092 Zürich
Switzerland

Phone: +41 (0)44 632 34 36
Email: info@ems-ph.org
Homepage: www.ems-ph.org

Cover picture: Black hole-powered jet of electrons and sub-atomic particles streams from center of galaxy M87. Courtesy NASA and The Hubble Heritage Team (STScI/AURA)

Typeset using the authors' TEX files: I. Zimmermann, Freiburg

Printed in Germany

Contents

General introduction vii

1 Introduction 1

2 The laws of General Relativity 9

2.1 The Einstein equations 9

2.1.1 Regular ellipticity and regular hyperbolicity 16

2.2 The Cauchy problem 19

2.2.1 Cauchy problem for the Einstein equations: local in time, existence and uniqueness of solutions 19

2.3 Decomposition of the Einstein equations 31

3 Asymptotic flatness at spacelike infinity and conserved quantities 35

3.1 Conserved quantities 35

3.2 Asymptotic flatness 54

3.2.1 The maximal time function 58

3.2.2 Positivity of the energy 60

3.3 Angular momentum 70

3.3.1 Independence from exhaustion 71

3.3.2 Conservation of angular momentum 71

3.4 The center of mass integrals 72

3.4.1 Conservation of center of mass integrals 76

4 The global stability of Minkowski spacetime 78

4.1 Statement of the problem 78

4.1.1 Field theories in a given spacetime 79

4.2 Sketch of the proof of the global stability of Minkowski spacetime . 103

4.2.1 The problem in General Relativity – two main difficulties . . . 103

4.2.2 Resolution of the first difficulty 103

4.2.3 Resolution of the second difficulty 109

4.2.4 The controlling quantity 124

4.2.5 The continuity argument 127

4.2.6 Estimates for the geometric quantities associated to the maximal foliation $\{\mathcal{H}_t\}$ 131

4.2.7 Estimates for the geometric quantities associated to the null foliation $\{C_u\}$ 132

4.2.8 Decomposition of a Weyl field with respect to the surfaces $S_{t,u}$ 137

4.2.9 The borderline error integrals 139

Bibliography 143

Index 145

General introduction

General Relativity is a theory proposed by Einstein in 1915 as a unified theory of space, time and gravitation. The theory's roots extend over almost the entire previous history of physics and mathematics.

Its immediate predecessor, Special Relativity, established in its final form by Minkowski in 1908, accomplished the unification of space and time in the geometry of a 4-dimensional affine manifold, a geometry of simplicity and perfection on par with that of the Euclidean geometry of space. The root of Special Relativity is Electromagnetic Theory, in particular Maxwell's incorporation of Optics, the theory of light, into Electrodynamics.

General Relativity is based on and extends Newton's theory of Gravitation as well as Newton's equations of motion. It is thus fundamentally rooted in Classical Mechanics.

Perhaps the most fundamental aspect of General Relativity however, is its geometric nature. The theory can be seen as a development of Riemannian geometry, itself an extension of Gauss' intrinsic theory of curved surfaces in Euclidean space.

The connection between gravitation and Riemannian geometry arose in Einstein's mind in his effort to uncover the meaning of what in Newtonian theory is the fortuitous equality of the inertial and the gravitational mass. Identification, via the equivalence principle, of the gravitational tidal force with spacetime curvature at once gave a physical interpretation of curvature of the spacetime manifold and also revealed the geometrical meaning of gravitation.

One sees here that descent to a deeper level of understanding of physical reality is connected with ascent to a higher level of mathematics. General Relativity constitutes a triumph of the geometric approach to physical science.

But there is more to General Relativity than merely a physical interpretation of a variant of Riemannian Geometry. For, the theory contains physical laws in the form of equations – Einstein's equations – imposed on the geometric structure. This gives a tightness which makes the resulting mathematical structure one of surpassing subtlety and beauty. An analogous situation is found by comparing the theory of differentiable functions of two real variables with the theory of differentiable functions of one complex variable. The latter gains, by the imposition of the Cauchy–Riemann equations, a tighter structure which leads to a greater richness of results.

The domain of application of General Relativity, beyond that of Newtonian theory, is astronomical systems, stellar or galactic, where the gravitational field is so strong that it implies the potential presence of velocities which are not negligible in comparison with the velocity of light. The ultimate domain of application is the study of the structure and evolution of the universe as a whole.

General Relativity has perhaps the most satisfying structure of all physical theories from the mathematical point of view. It is a wonderful research field for a mathematician. Here, results obtained by purely mathematical means have direct physical consequences.

One example of this is the incompleteness theorem of R. Penrose and its extensions due to Hawking and Penrose known as the "singularity theorems". This result is relevant to the study of the phenomenon of gravitational collapse. It shall be covered in the second volume of the present work. The methods used to establish the result are purely geometrical – the theory of conjugate points. In fact, part of the main argument is already present in the theory of focal points in the Euclidean framework, a theory developed in antiquity.

Another example is the positive energy theorem, the first proof of which, due to R. Schoen and S. T. Yau, is based on the theory of minimal surfaces and is covered in the the present volume. In this example a combination of geometric and analytic methods are employed.

A last example is the theory of gravitational radiation, a main theme for both volumes of this work. Here also we have a combination of geometric and analytic methods. A particular result in the theory of gravitational radiation is the so-called memory effect [11], which is due to the non-linear character of the asymptotic laws at future null infinity and has direct bearing on experiments planned for the near future. This result will also be covered in our second volume.

The laws of General Relativity, Einstein's equations, constitute, when written in any system of local coordinates, a non-linear system of partial differential equations for the metric components. Because of the compatibility conditions of the metric with the underlying manifold, when piecing together local solutions to obtain the global picture, it is the geometric manifold, namely the pair consisting of the manifold itself together with its metric, which is the real unknown in General Relativity.

The Einstein equations are of hyperbolic character, as is explained in detail in this first volume. As a consequence, the initial value problem is the natural mathematical problem for these equations. This conclusion, reached mathematically, agrees with what one expects physically. For, the initial value problem is the problem of determining the evolution of a system from given initial conditions, as in the prototype example of Newton's equations of motion. The initial conditions for Einstein's equations, the analogues of initial position and velocity of Newtonian mechanics, are the intrinsic geometry of the initial spacelike hypersurface and its rate of change under a virtual normal displacement, the second fundamental form. In contrast to the case of Newtonian mechanics however, these initial conditions are, by virtue of the Einstein equations themselves, subject to constraints, and it is part of the initial value problem in General Relativity – a preliminary part – to analyze these constraints. Important results can be obtained on the basis of this analysis alone and the positive energy theorem is an example of such a result.

An important notion in physics is that of an isolated system. In the context of the theory of gravitation, examples of such systems are a planet with its moons, a star with its planetary system, a binary or multiple star, a cluster of stars, a galaxy, a pair or multiplet of interacting galaxies, or, as an extreme example, a cluster of galaxies – but not the universe as a whole. What is common in these examples is that each of these systems can be thought of as having an asymptotic region in which conditions are trivial. Within General Relativity the trivial case is the flat Minkowski spacetime of Special Relativity. Thus the desire to describe isolated gravitating systems in General Relativity leads us to consider spacetimes with asymptotically Minkowskian regions. However it is important to remember at this point the point of view of the initial value problem: a spacetime is determined as a solution of the Einstein equations from its initial data. Consequently, we are not free to impose our own requirements on a spacetime. We are only free to impose requirements on the initial data – to the extent that the requirements are consistent with the constraint equations. Thus the correct notion of an isolated system in the context of General Relativity is a spacetime arising from asymptotically flat initial conditions, namely an intrinsic geometry which is asymptotically Euclidean and is a second fundamental form which tends to zero at infinity in an appropriate way. This is discussed in detail in this volume.

Trivial initial data for the Einstein equations consists of Euclidean intrinsic geometry and a vanishing second fundamental form. Trivial initial data gives rise to the trivial solution, namely the Minkowski spacetime. A natural question in the context of the initial value problem for the vacuum Einstein equations is whether or not every asymptotically flat initial data which is globally close to the trivial data gives rise to a solution which is a complete spacetime tending to the Minkowski spacetime at infinity along any geodesic. This question was answered in the affirmative in the joint work of the present author with Sergiu Klainerman, which appeared in the monograph [14]. One of the aims of the present work is to present the methods which went into that work in a more general context, so that the reader may more fully understand their origin and development as well as be able to apply them to other problems. In fact, problems coming from fields other than General Relativity are also treated in the present work. These fields are Continuum Mechanics, Electrodynamics of Continuous Media and Classical Gauge Theories (such as arise in the mesoscopic description of superfluidity and superconductivity). What is common to all these problems from our perspective is the mathematical methods involved.

One of the main mathematical methods analyzed and exploited in the present work is the general method of constructing a set of quantities whose growth can be controlled in terms of the quantities themselves. This method is an extension of the celebrated theorem of Noether, a theorem in the framework of the action principle, which associates a conserved quantity to each 1-parameter group of symmetries of the action (see [12]). This extension is involved at a most elementary level in the very definition of the notion of hyperbolicity for an Euler–Lagrange system of partial differential equations, as discussed in detail in this first volume. In fact we may say that such a system is hyperbolic at a particular background solution if linear perturbations about this solution possess positive energy in the high frequency limit.

The application of Noether's Principle to General Relativity requires the introduction of a background vacuum solution possessing a non-trivial isometry group, as is explained in this first volume. Taking Minkowski spacetime as the background, we have the symmetries of time translations, space translations, rotations and boosts, which give rise to the conservation laws of energy, linear momentum, angular momentum and center of mass integrals, respectively. However, as is explained in this first volume, these quantities have geometric significance only for spacetimes which are asymptotic at infinity to the background Minkowski spacetime, so that the symmetries are in fact asymptotic symmetries of the actual spacetime.

The other main mathematical method analyzed and exploited in the present work is the systematic use of characteristic (null) hypersurfaces. The geometry of null hypersurfaces has already been employed by R. Penrose in his incompleteness theorem mentioned above. What is involved in that theorem is the study of a neighborhood of a given null geodesic generator of such a hypersurface. On the other hand, in the work on the stability of Minkowski spacetime, the global geometry of a characteristic hypersurface comes into play. In addition, the properties of a foliation of spacetime by such hypersurfaces, also come into play. This method is used in conjunction with the first method, for, such characteristic foliations are used to define the actions of groups in spacetime which may be called quasi-conformal isometries, as they are globally as close as possible to conformal isometries and tend as rapidly as possible to conformal isometries at infinity. The method is introduced in this first volume and will be treated much more fully in the second volume. It has applications beyond General Relativity to problems in Fluid Mechanics and, more generally, to the Mechanics and Electrodynamics of Continuous Media.

This book is based on Nachdiplom Lectures held at the Eidgenössische Technische Hochschule Zurich during the Winter Semester 2002/2003. The author wishes to thank his former student Lydia Bieri for taking the notes of this lecture, from which a first draft was written, and for making the illustrations.

1 Introduction

The *general theory of relativity* is a unified theory of space, time and gravitation. The fundamental concept of the theory is the concept of a spacetime manifold.

Definition 1. A *spacetime manifold* is a 4-dimensional oriented differentiable manifold M, endowed with a Lorentzian metric g.

Definition 2. A *Lorentzian metric* g is a continuous assignment of a non-degenerate quadratic form g_p, of index 1, in T_pM at each $p \in M$.

Here we denote by T_pM the tangent space to M at p. Also, non-degenerate means

$$g_p(X, Y) = 0 \ \forall \ Y \in T_pM \implies X = 0,$$

while of index 1 means that the maximal dimension of a subspace of T_pM, on which g_p is negative definite, is 1.

An equivalent definition is the following.

Definition 3. A quadratic form g_p in T_pM is called Lorentzian if there exists a vector $V \in T_pM$ such that $g_p(V, V) < 0$ while setting

$$\Sigma_V = \{X : g_p(X, V) = 0\} \quad \text{(the “}g_p\text{-orthogonal complement of } V\text{”)},$$

$g_p|_{\Sigma_V}$ is positive definite.

We can then choose an orthonormal frame (E_0, E_1, E_2, E_3) at p, by setting

$$E_0 = \frac{V}{\sqrt{-g(V, V)}}$$

and choosing an orthonormal basis (E_1, E_2, E_3) for Σ_V. Given any vector $X \in T_pM$ we can expand

$$\begin{aligned} X &= X^0 E_0 + X^1 E_1 + X^2 E_2 + X^3 E_3 \\ &= \sum_\mu X^\mu E_\mu \quad (\mu = 0, 1, 2, 3). \end{aligned}$$

Then

$$\begin{aligned} g(E_\mu, E_\nu) &= \eta_{\mu\nu} = \operatorname{diag}(-1, 1, 1, 1), \\ g(X, X) &= -(X^0)^2 + (X^1)^2 + (X^2)^2 + (X^3)^2 \\ &= \sum_{\mu\nu} \eta_{\mu\nu} X^\mu X^\nu. \end{aligned}$$

Definition 4. The null cone at $p \in M$,

$$N_p = \{X \neq 0 \in T_pM \,:\, g_p(X,X) = 0\},$$

is a double cone $N_p = N_p^+ \cup N_p^-$.

Denote by I_p^+ the interior of N_p^+ and by I_p^- the interior of N_p^-.

Definition 5. The set of timelike vectors at $p \in M$ is defined as

$$I_p := I_p^+ \cup I_p^- = \{X \in T_pM \,:\, g_p(X,X) < 0\},$$

where I_p is an open set.

Definition 6. The set of spacelike vectors at $p \in M$ is defined as

$$S_p := \{X \in T_pM \,:\, g_p(X,X) > 0\},$$

where S_p is the exterior of N_p, a connected open set.

Time orientability. We assume that a continuous choice of positive (future) component I_p^+ of I_p at each $p \in M$, is possible. Once such a choice has been made, the spacetime manifold M is called *time oriented.*

Definition 7. A causal curve in M is a differentiable curve γ whose tangent vector $\dot{\gamma}$ at each point $p \in M$ belongs to $I_p \cup N_p$, that is, it is either timelike or null.

Then either $\dot{\gamma}_p \in I_p^+ \cup N_p^+$ at each p along γ in which case γ is called future-directed, or $\dot{\gamma}_p \in I_p^- \cup N_p^-$ at each p along γ in which case γ is called past-directed.

Given a point $p \in M$, we can then define the causal future of p.

Definition 8. The causal future of p, denoted by $J^+(p)$, is the set of all points $q \in M$ for which there exists a future-directed causal curve initiating at p and ending at q.

Similarly, we can define $J^-(p)$, the causal past of p.

Definition 9. The arc length of a causal curve γ between the points corresponding to the parameter values $\lambda = a,\ \lambda = b$ is

$$L[\gamma](a,b) = \int_a^b \sqrt{-g(\dot{\gamma}(\lambda), \dot{\gamma}(\lambda))}\, d\lambda.$$

If $q \in J^+(p)$, we define the temporal distance of q from p by

$$\tau(q,p) = \sup_{\substack{\text{all future-directed causal}\\ \text{curves from } p \text{ to } q}} L[\gamma].$$

The arc length is independent of the parametrization of the curve.

Let us recall the Hopf–Rinow theorem in Riemannian geometry:

Theorem 1 (Hopf–Rinow). *On a complete Riemannian manifold any two points can be joined by a minimizing geodesic.*

In Lorentzian geometry the analogous statement is in general false. It holds however when the spacetime admits a Cauchy hypersurface (the definition of this concept will follow). When the supremum in the above definition is achieved and the metric is C^1 the maximizing curve is a causal geodesic; after suitable reparametrization the tangent vector is parallelly transported along the curve.

Examples of spacetime manifolds. We take as our model the

- Riemannian spaces of constant curvature:

$$\frac{1}{(1+\frac{k}{4}|x|^2)^2}|dx|^2, \quad k=0,1,-1,$$

where $|\cdot|$ is the Euclidean magnitude, $|v|=\sqrt{\sum_i (v^i)^2}$. When $k=-1$ the manifold is the ball of radius 2 in $\mathbb{R}^n$, $|x|<2$.

By analogy, we have the

- Lorentzian spaces of constant curvature:

$$\frac{1}{(1+\frac{k}{4}\langle x,x\rangle)^2}\langle dx,dx\rangle.$$

Here $\langle\cdot,\cdot\rangle$ is the Minkowski quadratic form. $\langle u,v\rangle=-u^0v^0+\sum_{i=1}^{n-1}u^iv^i$. For $k=1$ we have what is called *de-Sitter-space* while for $k=-1$ we have what is called *Anti-de-Sitter-space*. In the case $k=-1$ the manifold is $\{x\in\mathbb{R}^n : \langle x,x\rangle<4\}$.

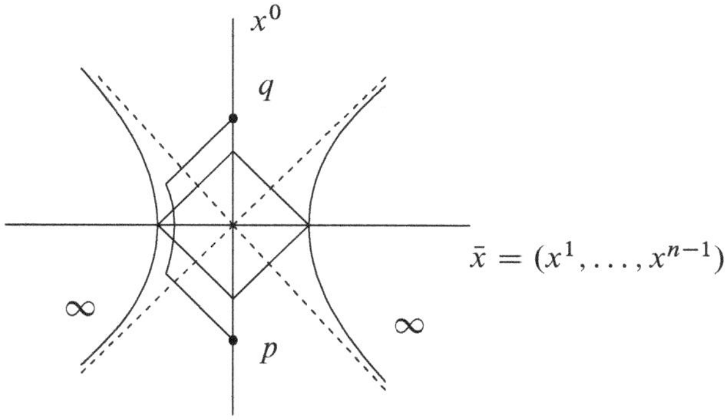

Figure 1

In Anti-de-Sitter space there are points p and q as shown for which $\tau(p,q) = \infty$. For, the length of the timelike segment of the causal curve joining p and q in the figure can be made arbitrarily large by making the segment approach the hyperboloid at infinity $\langle x, x\rangle = 4$.

Definition 10. We define the causal future $J^+(K)$ and the causal past $J^-(K)$ of any set $K \subset M$, in particular a closed set, by

$$J^{\pm}(K) = \{q \in M : q \in J^{\pm}(p) \text{ for some } p \in K\}.$$

The boundaries of $J^+(K)$, $J^-(K)$, i.e. $\partial J^+(K)$, $\partial J^-(K)$ for closed sets K, are null hypersurfaces. They are realized as level sets of functions u satisfying the eikonal equation $g^{\mu\nu}\partial_\mu u \partial_\nu u = 0$. These hypersurfaces are generated by null geodesic segments, as shall be shown below. They are thus analogous to ruled surfaces in Euclidean geometry. Moreover, the null geodesics generating $J^+(K)$ have past end-points only on K and those generating $J^-(K)$ have future end-points only on K. The null geodesics generating $J^+(K)$ may have future end-points, even when K is a single point. The set of these end-points forms the *future null cut locus* corresponding to K. Similarly for $J^-(K)$. (Null cut loci shall be discussed at length in the second volume.)

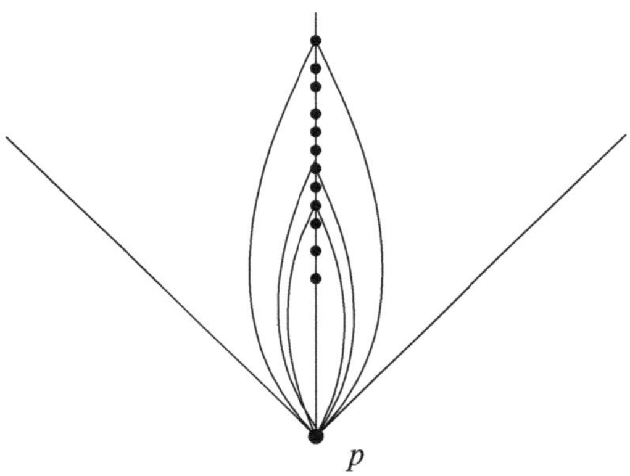

Figure 2. The future null cut locus of a point p. Each of the points marked by a dot is a point where a pair of null geodesics issuing from p intersect.

Definition 11. H is called a null hypersurface if at each point $x \in H$ the induced metric $g_x|_{T_x H}$ is degenerate.

Thus there exists an $L \neq 0 \in T_x H$ such that

$$g_x(L, X) = 0 \quad \forall\, X \in T_x H.$$

Now $T_x H$ is a hyperplane in $T_x M$. Such a hyperplane is defined by a covector $\xi \in T_x^* M$,

$$T_x H = \{X \in T_x M : \xi \cdot X = 0\}.$$

Representing H as the (0-)level set of a function u, we can take

$$\xi = du(x).$$

If we set $L^\mu = -g^{\mu\nu}\partial_\nu u$ (components with respect to an arbitrary frame), we have

$$g(L, X) = -du \cdot X$$

and L is g-orthogonal to H. Then H is a null hypersurface if and only if

$$L_x \in T_x H \quad \forall\, x \in H.$$

Now let u be a function, each of the level sets of which is a null hypersurface. Taking then $X = L$ we obtain

$$g(L, L) = 0 \quad (g_{\mu\nu} L^\mu L^\nu = 0),$$

hence L is at each point a null vector, a condition which, expressed in terms of du, reads

$$g^{\mu\nu}\partial_\mu u \partial_\nu u = 0,$$

which is the eikonal equation.

In fact, L is a geodesic vector field, that is, the integral curves of L are null geodesics. The proof of this fact is as follows.

$$(\nabla_L L)^\mu = L^\nu \nabla_\nu L^\mu, \tag{1}$$

$$g_{\mu\lambda}(\nabla_L L)^\mu = L^\nu \nabla_\nu L_\lambda, \tag{2}$$

where $L_\lambda = g_{\mu\lambda} L^\mu = -\partial_\lambda u$. Now the Hessian of a function is symmetric:

$$\nabla_\nu(\partial_\lambda u) = \nabla_\lambda(\partial_\nu u). \tag{3}$$

Thus,

$$\begin{aligned} L^\nu \nabla_\nu L_\lambda &= L^\nu \nabla_\lambda L_\nu \\ &= \partial_\lambda\Big(\frac{1}{2} g(L, L)\Big) = 0, \end{aligned} \tag{4}$$

that is,

$$\nabla_L L = 0. \tag{5}$$

Definition 12. A hypersurface H is called spacelike if at each $x \in H$, the induced metric

$$g_x|_{T_x H} =: \bar{g}_x$$

is positive definite.

Then $(H, \bar{g})$ is a proper Riemannian manifold. The g-orthogonal complement of $T_x H$ is a 1-dimensional subspace of $T_x M$ on which g_x is negative definite. Thus there exists a vector $N_x \in I_x^+$ of unit magnitude

$$g_x(N_x, N_x) = -1$$

whose span is this 1-dimensional subspace. We call N (the so-defined vector field along H) the future-directed unit normal to H.

Definition 13. The 2nd fundamental form k of H is a 2-covariant symmetric tensor field on H, or quadratic form in $T_x H$ at each $x \in H$, defined by

$$k(X, Y) = g(\nabla_X N, Y) \quad \forall\, X,\ Y \in T_x H. \tag{6}$$

Definition 14. A Cauchy hypersurface is a complete spacelike hypersurface H in M (i.e., $(H, \bar{g})$ is a complete Riemannian manifold) such that if γ is any causal curve through any point $p \in M$, then γ intersects H at exactly one point.

Examples of Cauchy hypersurfaces

- A *spacelike hyperplane* in Minkowski spacetime M is a Cauchy hypersurface for M.
- A *spacelike hyperboloid in Minkowski spacetime*,

$$-(x^0)^2 + \sum_{i=1}^{3} (x^i)^2 = -1, \quad x^0 > 0,$$

  is a complete Riemannian manifold (of constant negative curvature) but not a Cauchy hypersurface for M. It is however a Cauchy hypersurface for $I_0^+ \subset M$.
- The Anti-de-Sitter space does not admit a Cauchy hypersurface.

If we consider only future evolution the above definition is replaced by one in which γ is taken to be any *past-directed* causal curve. Then M is a manifold with boundary and H is the past boundary of M.

Definition 15. A spacetime admitting a Cauchy hypersurface is called globally hyperbolic.

Under the hypothesis of global hyperbolicity we can define a time function. This is a differentiable function t such that

$$dt \cdot X > 0 \quad \forall X \in I_p^+, \ \forall \ p \in M. \tag{7}$$

The manifold M is then diffeomorphic to the product $\mathbb{R}\times\overline{M}$ where $\overline{M}$ is a 3-manifold, each level set Σ_t of t being diffeomorphic to $\overline{M}$. $H = \Sigma_0$ is a Cauchy hypersurface.

Definition 16. The lapse function corresponding to a time function t is the function

$$\Phi = (-g^{\mu\nu}\partial_\mu t \partial_\nu t)^{-\frac{1}{2}}. \tag{8}$$

This measures the normal separation of the leaves Σ_t (of the foliation by the level sets of t). Consider the vector field

$$T^\mu = -\Phi^2 g^{\mu\nu}\partial_\nu t. \tag{9}$$

The integral curves of T are the orthogonal curves to the Σ_t-foliation. Moreover, $Tt = T^\mu\partial_\mu t = 1$. Thus the orthogonal curves are parametrized by t. That is, the 1-parameter group ϕ_τ generated by T takes the leaves onto each other: $\phi_\tau(\Sigma_t) = \Sigma_{t+\tau}$. Thus T is a time translation vector field. The unit normal N is given by

$$N = \Phi^{-1}T. \tag{10}$$

The integral curves of N are the same orthogonal curves but parametrized by arc length s. It follows that, along an orthogonal curve,

$$\frac{ds}{dt} = \Phi. \tag{11}$$

We can identify $\overline{M}$ with $\Sigma_0 = H$. The mapping of M into $\mathbb{R} \times \overline{M}$ or $[0,\infty) \times \overline{M}$, taking $p \in M$ to the pair (t, q) if p lies on Σ_t and along the orthogonal curve through $q \in \Sigma_0$, is a diffeomorphism.

In terms of this representation of M we can write

$$g = -\Phi^2 dt^2 + \bar{g}, \tag{12}$$

where $\bar{g} = \bar{g}(t)$ is the induced metric on Σ_t, which is positive definite. Moreover,

$$T = \frac{\partial}{\partial t}, \quad N = \frac{1}{\Phi}\frac{\partial}{\partial t}.$$

Assume that (E_1, E_2, E_3) is a frame field for Σ_t, which is Lie transported along (the integral curves of) T, that is,

$$[T, E_i] = 0, \quad i = 1, 2, 3. \tag{13}$$

(In particular we may take (x^1, x^2, x^3) to be local coordinates on $H = \Sigma_0$, and set $E_i = \frac{\partial}{\partial x^i}$.) Then we have

$$\begin{aligned}
k_{ij} = k(E_i, E_j) &= \frac{1}{2}(g(\nabla_{E_i} N, E_j) + g(\nabla_{E_j} N, E_i)) \\
&= \frac{1}{2\Phi}(g(\nabla_{E_i} T, E_j) + g(\nabla_{E_j} T, E_i)) \\
&= \frac{1}{2\Phi}(g(\nabla_T E_i, E_j) + g(\nabla_T E_j, E_i)) \\
&= \frac{1}{2\Phi} T(g(E_i, E_j));
\end{aligned}$$

that is

$$k_{ij} = \frac{1}{2\Phi}\frac{\partial \bar{g}_{ij}}{\partial t}, \tag{14}$$

where $\bar{g}_{ij} = \bar{g}(E_i, E_j) = g(E_i, E_j)$ are the components of the induced metric on Σ_t. The above equation is called the 1st variational formula.

2 The laws of General Relativity

2.1 The Einstein equations

The laws of General Relativity are the Einstein equations linking the curvature of spacetime to its matter content:

$$G_{\mu\nu} := R_{\mu\nu} - \frac{1}{2} g_{\mu\nu} R = 2T_{\mu\nu}. \tag{15}$$

(We are using rationalized units where 4π times Newton's gravitational constant as well as the speed of light in vacuum are set equal to 1.) Here $T_{\mu\nu}$ is the energy-momentum tensor of matter, $G_{\mu\nu}$ the Einstein tensor, $R_{\mu\nu}$ the Ricci tensor and R the scalar curvature of the metric $g_{\mu\nu}$. From the original Bianchi identity

$$\nabla_{[\alpha} R_{\beta\gamma]\delta\epsilon} := \nabla_\alpha R_{\beta\gamma\delta\epsilon} + \nabla_\beta R_{\gamma\alpha\delta\epsilon} + \nabla_\gamma R_{\alpha\beta\delta\epsilon} = 0, \tag{16}$$

one obtains

$$\nabla^\nu G_{\mu\nu} = 0, \tag{17}$$

the twice contracted Bianchi identity. This identity (17) implies

$$\nabla^\nu T_{\mu\nu} = 0, \tag{18}$$

the equations of motion of matter. The Einstein vacuum equations

$$G_{\mu\nu} = 0 \tag{19}$$

correspond to the absence of matter: $T_{\mu\nu} = 0$. The equations are then equivalent to

$$R_{\mu\nu} = 0. \tag{20}$$

The connection coefficients $\Gamma^\mu_{\alpha\beta}$ and the curvature and Ricci tensor components in arbitrary local coordinates read as follows:

$$\Gamma^\mu_{\alpha\beta} = \frac{1}{2} g^{\mu\nu} (\partial_\alpha g_{\beta\nu} + \partial_\beta g_{\alpha\nu} - \partial_\nu g_{\alpha\beta}), \tag{21}$$

$$R^\alpha_{\mu\lambda\nu} = \partial_\lambda \Gamma^\alpha_{\mu\nu} - \partial_\nu \Gamma^\alpha_{\mu\lambda} + \Gamma^\alpha_{\beta\lambda} \Gamma^\beta_{\mu\nu} - \Gamma^\alpha_{\beta\nu} \Gamma^\beta_{\mu\lambda}, \tag{22}$$

$$R_{\mu\nu} = R^\alpha_{\mu\alpha\nu} = \partial_\alpha \Gamma^\alpha_{\mu\nu} - \partial_\nu \Gamma^\alpha_{\mu\alpha} + \Gamma^\alpha_{\beta\alpha} \Gamma^\beta_{\mu\nu} - \Gamma^\alpha_{\beta\nu} \Gamma^\beta_{\mu\alpha}. \tag{23}$$

Denoting by P.P. the principal part, that is, the part containing the highest (2nd) derivatives of the metric, we have

$$\text{P.P.}\{R_{\mu\nu}\} = \frac{1}{2} g^{\alpha\beta} \{\partial_\mu \partial_\alpha g_{\beta\nu} + \partial_\nu \partial_\alpha g_{\beta\mu} - \partial_\mu \partial_\nu g_{\alpha\beta} - \partial_\alpha \partial_\beta g_{\mu\nu}\}. \tag{24}$$

We shall presently discuss the character of the Einstein equations as reflected in their symbol. The symbol is defined by replacing in the principal part

$$\partial_\mu \partial_\nu g_{\alpha\beta} \quad \text{by} \quad \xi_\mu \xi_\nu \dot{g}_{\alpha\beta},$$

where ξ_μ are the components of a covector and $\dot{g}_{\alpha\beta}$ the components of a possible variation of g. We then obtain the symbol σ_ξ at a point $p \in M$ and a covector $\xi \in T^*_p M$, for a given background metric g:

$$(\sigma_\xi \cdot \dot{g})_{\mu\nu} = \frac{1}{2} g^{\alpha\beta} (\xi_\mu \xi_\alpha \dot{g}_{\beta\nu} + \xi_\nu \xi_\alpha \dot{g}_{\beta\mu} - \xi_\mu \xi_\nu \dot{g}_{\alpha\beta} - \xi_\alpha \xi_\beta \dot{g}_{\mu\nu}).$$

Let us denote

$$\begin{aligned}
(i_\xi \dot{g})_\nu &= g^{\alpha\beta} \xi_\alpha \dot{g}_{\beta\nu}, \\
(\xi, \xi) &= g^{\alpha\beta} \xi_\alpha \xi_\beta, \\
(\xi \otimes \zeta)_{\mu\nu} &= \xi_\mu \zeta_\nu, \\
g^{\alpha\beta} \dot{g}_{\alpha\beta} &= \operatorname{tr} \dot{g}.
\end{aligned}$$

We can then write

$$(\sigma_\xi \cdot \dot{g}) = \frac{1}{2} \{\xi \otimes i_\xi \dot{g} + i_\xi \dot{g} \otimes \xi - \operatorname{tr} \dot{g} \xi \otimes \xi - (\xi, \xi) \dot{g}\}.$$

The notion of the symbol of a system of Euler–Lagrange equations is as follows. Let us denote by x, the independent variables: x^μ, $\mu = 1, \dots, n$; by q, the dependent variables: $q^a, a = 1, \dots, m$; and by v, the 1st derivatives of dependent variables: v^a_μ, $n \times m$ matrices. Then the Lagrangian L is a given function of (x, q, v),

$$L = L(x, q, v).$$

A set of functions $(u^a(x) : a = 1, \dots, m)$ is a *solution of the Euler–Lagrange equations*, if substituting

$$\begin{aligned}
q^a &= u^a(x), \\
v^a_\mu &= \frac{\partial u^a}{\partial x^\mu}(x)
\end{aligned}$$

we have

$$\frac{\partial}{\partial x^\mu} \Big(\frac{\partial L}{\partial v^a_\mu}(x, u(x), \partial u(x)) \Big) - \frac{\partial L}{\partial q^a}(x, u(x), \partial u(x)) = 0. \tag{25}$$

Defining

$$\begin{aligned}
p^\mu_a &= \frac{\partial L}{\partial v^a_\mu}, \\
f_a &= \frac{\partial L}{\partial q^a},
\end{aligned}$$

the Euler–Lagrange equations become

$$\frac{\partial p_a^\mu}{\partial x^\mu} = f_a.$$

The q, v, p, f are analogous to position, velocity, momentum and force, respectively, in classical mechanics.

The principal part of the Euler–Lagrange equations is

$$h_{ab}^{\mu\nu} \frac{\partial^2 u^b}{\partial x^\mu \partial x^\nu}(x, u(x), \partial u(x)),$$

where

$$h_{ab}^{\mu\nu} = \frac{\partial^2 L}{\partial v_\mu^a \partial v_\nu^b}(x, q, v).$$

Let us consider the equations of variation. These are the linearized equations, satisfied by a variation through solutions. If we denote by $\dot{u}^a$ the variations of the functions u^a, the principal part of the linearized equations is

$$h_{ab}^{\mu\nu}(x, u(x), \partial u(x)) \frac{\partial^2 \dot{u}^b}{\partial x^\mu \partial x^\nu}.$$

Consider in fact oscillatory solutions

$$\dot{u}^a = \dot{w}^a e^{i\Phi} \tag{26}$$

of the equations of variation. Writing $\frac{\Phi}{\epsilon}$ in place of Φ, substituting in the linearized equations and keeping only the leading terms as $\epsilon \to 0$ (high frequency limit), we obtain

$$h_{ab}^{\mu\nu}(x, u(x), \partial u(x))\, \dot{w}^b \frac{\partial \Phi}{\partial x^\mu} \frac{\partial \Phi}{\partial x^\nu} = 0. \tag{27}$$

The left-hand side is the symbol $\sigma_\xi \cdot \dot{w}$, where $\xi_\mu = \frac{\partial \Phi}{\partial x^\mu}$. Thus, the symbol of the Euler–Lagrange equations is in general given by

$$(\sigma_\xi \cdot \dot{u})^a = h_{ab}^{\mu\nu} \xi_\mu \xi_\nu \dot{u}^b = \chi_{ab}(\xi) \dot{u}^b, \tag{28}$$

where

$$\chi_{ab}(\xi) = h_{ab}^{\mu\nu} \xi_\mu \xi_\nu \tag{29}$$

is an $m \times m$ matrix whose entries are homogeneous quadratic polynomials in ξ.

From a global perspective, the x^μ, $\mu = 1, \dots, n$ are local coordinates on an n-dimensional manifold M and x denotes an arbitrary point on M, while the q^a, $a = 1, \dots, m$ are local coordinates on an m-dimensional manifold N and q denotes an arbitrary point on N. The unknown u is then a mapping $u \colon M \to N$ and the functions $(u^a(x), a = 1, \dots, m)$ describe this mapping in terms of the given local coordinates.

Definition 17. Let M be an n-dimensional manifold. Then the characteristic subset $C_x^* \subset T_x^* M$ is defined by

$$\begin{aligned} C_x^* &= \{\xi \neq 0 \in T_x^* M : \text{ null space of } \sigma_\xi \neq 0\} \\ &= \{\xi \neq 0 \in T_x^* M : \det \chi(\xi) = 0\}. \end{aligned}$$

Thus $\xi \in C_x^*$ if and only if $\xi \neq 0$ and the null space of σ_ξ is non-trivial.

The simplest example of an Euler–Lagrange equation with a non-empty characteristic is the linear wave equation

$$\Box u := g^{\mu\nu} \nabla_\mu (\partial_\nu u) = 0.$$

This equation arises from the Lagrangian

$$L = \frac{1}{2} g^{\mu\nu} v_\mu v_\nu.$$

The symbol is $\sigma_\xi \cdot \dot{u} = (g^{\mu\nu} \xi_\mu \xi_\nu) \dot{u}$ and the characteristic is

$$C_x^* = \{\, \xi \neq 0 \in T_x^* M : \ (\xi, \xi) = g^{\mu\nu} \xi_\mu \xi_\nu = 0\},$$

that is, C_x^* is the null cone in $T_x^* M$ associated to the metric g.

Let us now return to the symbol for the Einstein equations. Let us set

$$\dot{g} = \zeta \otimes \xi + \xi \otimes \zeta \tag{30}$$

for an arbitrary covector $\zeta \in T_x^* M$. Then

$$i_\xi \dot{g} = \underbrace{(\zeta, \xi)}_{= g^{\mu\nu} \zeta_\mu \xi_\nu} \ \xi + (\xi, \xi)\, \zeta \tag{31}$$

and

$$\operatorname{tr} \dot{g} = 2(\zeta, \xi). \tag{32}$$

We see that

$$\sigma_\xi \cdot \dot{g} = 0. \tag{33}$$

Therefore the null space of σ_ξ is non-trivial for every covector ξ. This degeneracy is due to the fact that the equations are generally covariant. That is, if g is a solution of the Einstein equations and f is a diffeomorphism of the manifold onto itself, then the pullback $f^* g$ is also a solution. If X is a (complete) vector field on M, then X generates a 1-parameter group $\{f_t\}$ of diffeomorphisms of M and

$$\mathcal{L}_X g = \frac{d}{dt} f_t^* g \Big|_{t=0}, \tag{34}$$

the Lie derivative with respect to X of g, is a solution of the linearized equations.

Let us recall that the Lie derivative of g with respect to a vector field X is given by

$$(\mathcal{L}_X g)(Y, Z) = g(\nabla_Y X, Z) + g(\nabla_Z X, Y).$$

Setting $Y = E_\mu$ and $Z = E_\nu$, where $(E_\mu;\ \mu = 0, \dots, 3)$ is an arbitrary frame, we can write

$$(\mathcal{L}_X g)_{\mu\nu} = \nabla_\mu X_\nu + \nabla_\nu X_\mu,$$

where $X_\mu = g_{\mu\lambda} X^\lambda$. The symbol of a Lie derivative is given by

$$\dot{g}_{\mu\nu} = \xi_\mu \zeta_\nu + \xi_\nu \zeta_\mu, \quad \text{where } \zeta_\mu = \dot{X}_\mu.$$

A simple analogue. The *Maxwell equations* for the electromagnetic field $F_{\mu\nu}$,

$$\nabla^\nu F_{\mu\nu} = g^{\nu\lambda} \nabla_\lambda F_{\mu\nu} = 0, \tag{35}$$

provide a simple analogue to this situation. Let us recall that $F = dA$, or $F_{\mu\nu} = \partial_\mu A_\nu - \partial_\nu A_\mu$, where A_μ is the electromagnetic potential, a 1-form. The Maxwell equations are the Euler–Lagrange equations of the Lagrangian

$$L = \frac{1}{4} F^{\mu\nu} F_{\mu\nu}$$

where $F^{\mu\nu} = g^{\mu\kappa} g^{\nu\lambda} F_{\kappa\lambda}$. The symbol for these equations is

$$(\sigma_\xi \cdot \dot{A})_\mu = g^{\nu\lambda} \xi_\lambda (\xi_\mu \dot{A}_\nu - \xi_\nu \dot{A}_\mu),$$

that is

$$\sigma_\xi \cdot \dot{A} = (\xi, \dot{A})\xi - (\xi, \xi)\dot{A}.$$

Consider the variation

$$\dot{A} = \lambda\, \xi$$

for any real number λ. Then

$$\sigma_\xi \cdot \dot{A} = 0.$$

Thus we have a degeneracy here as well: the null space of σ_ξ is non-trivial for all $\xi \in T_x^* M$. This is due to the gauge invariance of the Maxwell equations. If A is a solution, so is

$$\tilde{A} = A + df$$

for any function f. In fact $\tilde{A}$ is considered to be equivalent to A, just as $f^* g$ is considered to be equivalent to g. Thus (by linearity) $\dot{A}_\mu = \partial_\mu f$ is a solution of the linearized equations, for any function f. To remove the degeneracy we must factor out these trivial solutions. Correspondingly in General Relativity we must factor out the solutions of the form $\mathcal{L}_X g = \dot{g}$ for any vector field X. At the level of the symbol the gauge transformation is

$$\tilde{\dot{A}}_\mu = \dot{A}_\mu + \xi_\mu \dot{f}.$$

So, we introduce the equivalence relation

$$\dot{A}_1 \sim \dot{A}_2 \iff \dot{A}_2 = \dot{A}_1 + \lambda\xi, \ \lambda \in \mathbb{R}.$$

We then obtain a quotient space Q of dimension $4 - 1 = 3$. Consider now the null space of σ_ξ with σ_ξ defined on Q. We distinguish two cases: in the first case $(\xi, \xi) \neq 0$ and in the second case $(\xi, \xi) = 0$.

Case 1. $(\xi, \xi) \neq 0$. Then

$$\sigma_\xi \cdot \dot{A} = 0 \implies \dot{A} = \lambda\xi, \ \lambda = \frac{(\xi, \dot{A})}{(\xi, \xi)},$$

that is $\dot{A} \sim 0$. Thus we have the trivial null space if ξ is not a null covector.

Case 2. $(\xi, \xi) = 0$. In this case we may choose another covector $\underline{\xi}$ in the same component of the null cone such that $(\xi, \underline{\xi}) = -2$. There is then a unique representative $\dot{A}$ in each equivalence class in Q such that

$$(\underline{\xi}, \dot{A}) = 0.$$

For, take another element $\dot{A}'$ out of the equivalence class of $\dot{A}$, that is $\dot{A}' = \dot{A} + \lambda\xi$ for some $\lambda \in \mathbb{R}$. Then

$$0 = (\underline{\xi}, \dot{A}') = (\underline{\xi}, \dot{A}) - 2\lambda$$

implies that $\dot{A}$ is the unique representative of its equivalence class with $(\underline{\xi}, \dot{A}) = 0$. Let us work with this representation. Then it holds that

$$\sigma_\xi \cdot \dot{A} = (\xi, \dot{A})\xi = 0, \ \xi \neq 0 \iff (\xi, \dot{A}) = 0.$$

We conclude that the null space of σ_ξ consists of the spacelike 2-dimensional plane Π, the g-orthogonal complement of the timelike plane spanned by ξ and $\underline{\xi}$. So, Π is the space of the degrees of freedom of the electromagnetic field at a point (two polarizations).

Returning to the symbol for the Einstein equations, the symbol for the Lie derivative

$$(\mathcal{L}_X g)_{\mu\nu} = \nabla_\mu X_\nu + \nabla_\nu X_\mu$$

reads, as noted above,

$$\xi_\mu \dot{X}_\nu + \xi_\nu \dot{X}_\mu,$$

where the $\dot{X}_\mu$ are the components of an arbitrary covector. Consider then the equivalence relation

$$\dot{g}_1 \sim \dot{g}_2 \iff \dot{g}_2 = \dot{g}_1 + \zeta \otimes \xi + \xi \otimes \zeta, \quad \zeta \in T_x^* M,$$

which gives a quotient space Q.

Again we distinguish the two cases according as to whether the covector ξ satisfies $(\xi, \xi) \neq 0$ or $(\xi, \xi) = 0$.

Case 1. $(\xi, \xi) \neq 0$. If ξ is *not* null, then $\sigma_\xi \cdot \dot{g} = 0$ implies that

$$\dot{g} = \zeta \otimes \xi + \xi \otimes \zeta, \quad \text{where } \zeta = \frac{(i_\xi \dot{g} - \frac{1}{2} \operatorname{tr} \dot{g} \xi)}{(\xi, \xi)},$$

thus σ_ξ has only trivial null space on Q.

Case 2. $(\xi, \xi) = 0$. If ξ *is* null, we can choose $\underline{\xi}$ in the same component of the null cone N_x^* in $T_x^* M$ such that $(\xi, \underline{\xi}) = -2$. There is then a unique representative $\dot{g}$ in each equivalence class $\{\dot{g}\} \in Q$ such that

$$i_{\underline{\xi}} \dot{g} = 0.$$

So,

$$\sigma_\xi \cdot \dot{g} = 0 \iff \xi \otimes i_\xi \dot{g} + i_\xi \dot{g} \otimes \xi - \xi \otimes \xi \operatorname{tr} \dot{g} = 0.$$

Taking the inner product with $\underline{\xi}$ we see that $(i_\xi \dot{g}, \underline{\xi}) = (i_{\underline{\xi}} \dot{g}, \xi) = 0$, hence

$$-2\, i_\xi \dot{g} + 2\, \xi \operatorname{tr} \dot{g} = 0.$$

Taking again the inner product with $\underline{\xi}$ gives

$$-4 \operatorname{tr} \dot{g} = 0, \quad \text{that is,} \quad \operatorname{tr} \dot{g} = 0.$$

Substituting this above yields

$$i_\xi \dot{g} = 0.$$

Conversely, $i_{\underline{\xi}} \dot{g} = i_\xi \dot{g} = 0$ and $\operatorname{tr} \dot{g} = 0$ implies that $\dot{g}$ lies in the null space of σ_ξ. Therefore, if $\xi \in N_x^*$, then the null space of σ_ξ can be identified with the space of trace-free quadratic forms on the 2-dimensional spacelike plane Π, the g-orthogonal complement of the linear span of ξ and $\underline{\xi}$. This is the space of gravitational degrees of freedom at a point (two polarizations).

2.1.1 Regular ellipticity and regular hyperbolicity. We have already introduced the quadratic form

$$h^{\mu\nu}_{ab} = \frac{\partial^2 L}{\partial v^a_\mu \partial v^b_\nu}(x, q, v)$$

in the general context of a Lagrangian theory of mappings $u\colon M \to \mathcal{N}$. A point $x \in M$ is represented in terms of local coordinates in M by $(x^\mu : \mu = 1, \dots, n)$. The position q is a possible value of $u(x)$, that is, a point in $\mathcal{N}$, represented by $(q^a : a = 1, \dots, m)$ in terms of local coordinates in $\mathcal{N}$, while the velocity v is a possible value of $du(x)$ and is represented by the $n \times m$-matrix $v^a_\mu = \frac{\partial u^a}{\partial x^\mu}(x)$. Here $n = \dim M$ and $m = \dim \mathcal{N}$.

Before stating the definition of regular ellipticity let us have a closer look at the necessary notions. Let $u\colon M \to \mathcal{N}$ be a background solution, $x \in M$, $q = u(x)$, and let $\xi \in T^*_x M$, $Q \in T_q\mathcal{N}$. Then Q is a variation in position, the value at x of a possible variation $\dot{u}$ of the background solution. The corresponding variation in velocity $\dot{v}$ is a linear map from $T_x M$ to $T_q\mathcal{N}$: $\dot{v} \in \mathcal{L}(T_x M, T_q\mathcal{N})$. For any $X \in T_x M$ the components of the vector $Q = \dot{v} \cdot X \in T_q\mathcal{N}$ are $Q^a = \dot{v}^a_\mu X^\mu$, where the X^μ are the components of X and the $\dot{v}^a_\mu$ are the components of $\dot{v}$. The space $S_2(\mathcal{L}(T_x M, T_q\mathcal{N}))$ of quadratic forms on $\mathcal{L}(T_x M, T_q\mathcal{N})$ splits into the direct sum

$$S_2 = S_{2+} \oplus S_{2-},$$

where S_{2+} consists of the even quadratic forms and S_{2-} of the odd quadratic forms. Thus, a quadratic form h on $\mathcal{L}(T_x M, T_q\mathcal{N})$ decomposes into

$$h = h_+ + h_-,$$

where h_+ and h_- are, respectively, the even and odd parts of h. In terms of components we have

$$h^{\mu\nu}_{ab} = h^{\mu\nu}_{+ab} + h^{\mu\nu}_{-ab},$$

where

$$h^{\nu\mu}_{ba} = h^{\mu\nu}_{ab}$$

(h being a symmetric bilinear form), and

$$\begin{aligned} h^{\nu\mu}_{+ab} &= h^{\mu\nu}_{+ba} = h^{\mu\nu}_{+ab}, \\ h^{\nu\mu}_{-ab} &= h^{\mu\nu}_{-ba} = -h^{\mu\nu}_{-ab}. \end{aligned}$$

We also need the following notion:

Definition 18. Rank-1-elements of $\mathcal{L}(T_x M, T_q\mathcal{N})$ are the elements $\dot{v}$ of the form

$$\dot{v} = \xi \otimes Q, \quad \xi \in T^*_x M,\ Q \in T_q\mathcal{N},$$

that is, $\dot{v} \cdot X = (\xi \cdot X)Q$ for all X in $T_x M$.

Now consider the quadratic form $h(\dot{v}, \dot{v}) = h^{\mu\nu}_{ab} \dot{v}^a_\mu \dot{v}^b_\nu$ on the velocity variations $\dot{v}^a_\mu$.

Regular ellipticity (Legendre–Hadamard condition). A Lagrangian L is called *regularly elliptic at* (x, q, v) if the quadratic form $h = \frac{\partial^2 L}{\partial v^2}(x, q, v)$ on $\mathcal{L}(T_x M, T_q \mathcal{N})$ is positive definite on the set of rank-1-elements $\dot{v}^a_\mu = \xi_\mu Q^a$ with $\xi \in T^*_x M$ and $Q \in T_q \mathcal{N}$.

If L and L' are two Lagrangians giving rise to the same Euler–Lagrange equations, then the difference $h - h'$ of the corresponding quadratic forms is an odd quadratic form.

Remark. The definition of regular ellipticity is independent of the choice of Lagrangian for the same Euler–Lagrange equations because odd quadratic forms vanish on the set of rank-1-elements.

Next we define regular hyperbolicity, a notion expounded in [12].

Definition 19. A Lagrangian L is called regularly hyperbolic at (x, q, v) if the quadratic form $h = \frac{\partial^2 L}{\partial v^2}(x, q, v)$ on $\mathcal{L}(T_x M, T_q \mathcal{N})$ has the following property: There exists a pair (ξ, X) in $T^*_x M \times T_x M$ with $\xi \cdot X > 0$ such that:

1. h is negative definite on the space
$$L_\xi = \{\xi \otimes Q : \quad Q \in T_q \mathcal{N}\},$$
2. h is positive definite on the set of rank-1-elements of the subspace
$$\Sigma_X = \{\dot{v} \in \mathcal{L}(T_x M, T_q \mathcal{N}) : \quad \dot{v} \cdot X = 0\}.$$

Note that this definition is also independent of the choice of Lagrangian giving rise to the same Euler–Lagrange equations.

Definition 20. Given a quadratic form h on $\mathcal{L}(T_x M, T_q \mathcal{N})$ and a pair (ξ, X) in $T^*_x M \times T_x M$ with $\xi \cdot X > 0$, we define a new quadratic form $m(\xi, X)$ on $\mathcal{L}(T_x M, T_q \mathcal{N})$ depending linearly on ξ and X by

$$m(\xi, X)(\dot{v}_1, \dot{v}_2) = (\xi \cdot X) h(\dot{v}_1, \dot{v}_2) - h(\xi \otimes \dot{v}_1 \cdot X, \dot{v}_2) - h(\dot{v}_1, \xi \otimes \dot{v}_2 \cdot X). \quad (36)$$

We call this the Noether transform of h defined by (ξ, X).

Proposition 1. *A Lagrangian L is regularly hyperbolic at (x, q, v) if and only if there exists a pair (ξ, X) in $T^*_x M \times T_x M$ with $\xi \cdot X > 0$ such that the Noether transform $m(\xi, X)$ of h corresponding to (ξ, X) is positive definite on the set*

$$R_\xi = \{\xi \otimes P + \zeta \otimes Q : \ \forall\, \zeta \in T^*_x M, \ \forall\, P, Q \in T_q \mathcal{N}\}$$

(*which is a set of special rank-2-elements*).

Remark. If h is an odd quadratic form, then the Noether transform of h corresponding to (ξ, X) vanishes on R_ξ.

Given h and $\xi \neq 0 \in T_x^* M$, we define $\chi(\xi)$, a quadratic form in $T_q \mathcal{N}$, by

$$\chi(\xi)(Q_1, Q_2) = h(\xi \otimes Q_1, \xi \otimes Q_2). \tag{37}$$

That is,

$$\chi_{ab}(\xi) = h_{ab}^{\mu\nu} \xi_\mu \xi_\nu.$$

Then the characteristic subset C_x^* of $T_x^* M$ is defined by

$$C_x^* = \{\, \xi \neq 0 \in T_x^* M : \chi(\xi) \text{ is singular} \,\}. \tag{38}$$

Also, given $Q \neq 0 \in T_q \mathcal{N}$, we define a quadratic form $\Psi(Q)$ in $T_x^* M$ as follows:

$$\Psi(Q)(\xi_1, \xi_2) = h(\xi_1 \otimes Q, \xi_2 \otimes Q), \tag{39}$$

that is,

$$\Psi^{\mu\nu}(Q) = h_{ab}^{\mu\nu} Q^a Q^b.$$

Next, for a given $\xi \in C_x^*$ we define

$$\begin{aligned} \Lambda(\xi) &= \{\, \Psi(Q) \cdot \xi : Q \neq 0 \in \text{null space of } \chi(\xi) \} \\ &\subset \Sigma_\xi = \{\, X \in T_x M : \xi \cdot X = 0 \} \quad \text{(a hyperplane in } T_x M). \end{aligned} \tag{40}$$

Here $\Psi(Q)$ is considered as a linear map of $T_x^* M$ into $T_x M$,

$$\xi_\mu \mapsto \Psi^{\mu\nu}(Q) \xi_\nu.$$

$\Lambda(\xi)$ is a positive cone in Σ_ξ. That is, if $X \in \Lambda(\xi)$ and $\lambda > 0$, then λX lies in $\Lambda(\xi)$. Also the following holds:

$$\Lambda(\mu \xi) = \Lambda(\xi) \quad \forall\, \mu > 0.$$

If ξ is a regular point of C_x^*, then the null space of $\chi(\xi)$ has dimension 1 and $\Lambda(\xi)$ is a ray. Otherwise the maximal dimension of $\Lambda(\xi)$ is $\dim \Sigma_\xi = n - 1$.

Definition 21. The characteristic subset C_x of $T_x M$ is defined by

$$C_x = \bigcup_{\xi \in C_x^*} \Lambda(\xi). \tag{41}$$

2.2 The Cauchy problem

2.2.1 Cauchy problem for the Einstein equations: local in time, existence and uniqueness of solutions. In this chapter we shall discuss the work of Y. Choquet-Bruhat in [8]. This work is based on the reduction of the Einstein equations to wave equations. To accomplish this reduction one has to introduce harmonic, or wave, coordinates.

Definition 22. Let (M, g) be a Riemannian manifold. Then a function Φ is called *harmonic* if

$$\triangle_g \Phi = 0, \tag{42}$$

where $\triangle_g \Phi = g^{\mu\nu}\, \nabla_\mu(\partial_\nu \Phi)$.

If the metric g is Lorentzian, then the equation $\triangle_g \Phi = 0$ is the *wave equation.*

Now the problem is the following: Given a coordinate chart (U, x) with $x = (x^0, x^1, x^2, x^3)$, find functions Φ^μ, $\mu = 0, 1, 2, 3$, each of which is a solution of the wave equation in U,

$$\triangle_g \Phi = 0 \quad \text{in U},$$

and such that, setting

$$\bar{x}^\mu = \Phi^\mu(x^0, x^1, x^2, x^3),$$

we have a diffeomorphism of the range V in $\mathbb{R}^4$ of the given chart onto another domain $\bar{V}$ in $\mathbb{R}^4$.

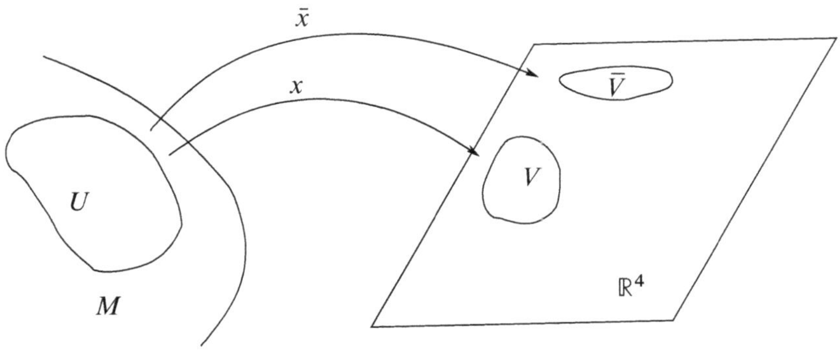

Figure 3

We thus have another chart $(U, \bar{x})$ with domain U, i.e. another system of local coordinates in U. The equation $\triangle_g \Phi = 0$ in an arbitrary system of local coordinates reads

$$\triangle_g \Phi = g^{\mu\nu} \left(\frac{\partial^2 \Phi}{\partial x^\mu \partial x^\nu} - \Gamma^\alpha_{\mu\nu} \frac{\partial \Phi}{\partial x^\alpha} \right) = 0. \tag{43}$$

Suppose now that we use the functions $(\bar{x}^0, \bar{x}^1, \bar{x}^2, \bar{x}^3)$ as the local coordinates in U, i.e. we express things relative to the new chart $(U, \bar{x})$. Setting Φ equal to each one of the $\bar{x}^\beta$ for $\beta = 0, 1, 2, 3$, we have a solution of the above equation. Since

$$\frac{\partial \bar{x}^\beta}{\partial \bar{x}^\gamma} = \delta^\beta_\gamma$$

we have

$$\frac{\partial^2 \bar{x}^\beta}{\partial \bar{x}^\mu \partial \bar{x}^\nu} = 0.$$

So, the equation reads

$$0 = \triangle_g \bar{x}^\beta = -\bar{g}^{\mu\nu} \bar{\Gamma}^\beta_{\mu\nu}.$$

Dropping the bars we can say that a system of local coordinates is harmonic if and only if the connection coefficients in these coordinates satisfy the condition

$$\Gamma^\alpha := g^{\mu\nu} \Gamma^\alpha_{\mu\nu} = 0. \tag{44}$$

Let us set

$$\Gamma_\alpha := g_{\alpha\beta} \Gamma^\beta.$$

Then we can write

$$\Gamma_\mu = g^{\alpha\beta} \partial_\alpha g_{\beta\mu} - \frac{1}{2} g^{\alpha\beta} \partial_\mu g_{\alpha\beta}.$$

Hence the principal part of $\partial_\mu \Gamma_\nu + \partial_\nu \Gamma_\mu$ is the following:

$$\text{P. P.}\{\partial_\mu \Gamma_\nu + \partial_\nu \Gamma_\mu\} = g^{\alpha\beta} \{\partial_\alpha \partial_\mu g_{\beta\nu} + \partial_\alpha \partial_\nu g_{\beta\mu} - \partial_\mu \partial_\nu g_{\alpha\beta}\}.$$

Denoting by $R_{\mu\nu}$ the components of the Ricci curvature tensor, let us define

$$H_{\mu\nu} = R_{\mu\nu} - \frac{1}{2}(\partial_\mu \Gamma_\nu + \partial_\nu \Gamma_\mu). \tag{45}$$

Then the principal part of $H_{\mu\nu}$ is

$$\text{P. P.}\{H_{\mu\nu}\} = -\frac{1}{2} g^{\alpha\beta} \partial_\alpha \partial_\beta g_{\mu\nu},$$

and we have

$$H_{\mu\nu} = -\frac{1}{2} g^{\alpha\beta} \partial_\alpha \partial_\beta g_{\mu\nu} + B^{\alpha\beta\kappa\lambda\rho\sigma}_{\mu\nu} \partial_\alpha g_{\kappa\lambda} \partial_\beta g_{\rho\sigma}, \tag{46}$$

where B is a rational function of the metric g of degree -2, the ratio of a homogeneous polynomial in g of degree 6 to $(\det g)^2$. Replacing the Einstein equations

$$R_{\mu\nu} = 0 \tag{47}$$

by the *reduced* (Einstein) equations

$$H_{\mu\nu} = 0, \tag{48}$$

we have a system of non-linear wave equations for the metric components $g_{\mu\nu}$.

In the approach of Choquet-Bruhat one studies the Cauchy problem for these reduced equations. Let us write

$$R_{\mu\nu} = H_{\mu\nu} + \frac{1}{2} S_{\mu\nu}, \tag{49}$$

where

$$S_{\mu\nu} = \partial_\mu \Gamma_\nu + \partial_\nu \Gamma_\mu .$$

We have

$$R_{\mu\nu} - \frac{1}{2} g_{\mu\nu} R = H_{\mu\nu} - \frac{1}{2} g_{\mu\nu} H + \frac{1}{2}\Big(S_{\mu\nu} - \frac{1}{2} g_{\mu\nu} S \Big),$$

and

$$\begin{aligned} \nabla^\nu \underbrace{(S_{\mu\nu} - \frac{1}{2} g_{\mu\nu} S)}_{=:\hat{S}_{\mu\nu}} &= g^{\nu\lambda} \nabla_\lambda \hat{S}_{\mu\nu} && (50) \\ &= g^{\nu\lambda} (\partial_\lambda \hat{S}_{\mu\nu} - \Gamma^\kappa_{\lambda\mu} \hat{S}_{\kappa\nu} - \Gamma^\kappa_{\lambda\nu} \hat{S}_{\mu\kappa}). && (51) \end{aligned}$$

If we have a solution of the reduced equations, then by virtue of the twice contracted Bianchi identities

$$\nabla^\nu \Big(R_{\mu\nu} - \frac{1}{2} g_{\mu\nu} R \Big) = 0, \tag{52}$$

this solution also satisfies the equations

$$\nabla^\nu \hat{S}_{\mu\nu} = 0. \tag{53}$$

Now, we have $S = 2\partial^\nu \Gamma_\nu$ with $\partial^\nu = g^{\nu\lambda} \partial_\lambda$. Therefore

$$\hat{S}_{\mu\nu} = \partial_\mu \Gamma_\nu + \partial_\nu \Gamma_\mu - g_{\mu\nu} \partial^\lambda \Gamma_\lambda .$$

The principal part of $\nabla^\nu \hat{S}_{\mu\nu}$ is

$$\begin{aligned} \mathrm{P.P.}\{\nabla^\nu \hat{S}_{\mu\nu}\} &= \partial_\mu (\partial^\nu \Gamma_\nu) + \partial^\nu \partial_\nu \Gamma_\mu - \partial_\mu (\partial^\lambda \Gamma_\lambda) \\ &= g^{\alpha\beta} \partial_\alpha \partial_\beta \Gamma_\mu . \end{aligned}$$

In fact, we have

$$\nabla^\nu \hat{S}_{\mu\nu} = g^{\alpha\beta} \partial_\alpha \partial_\beta \Gamma_\mu + A^{\alpha\beta}_\mu \partial_\alpha \Gamma_\beta ,$$

where A is a linear form in ∂g with coefficients which are homogeneous rational functions of g. Therefore the equations

$$\nabla^\nu \hat{S}_{\mu\nu} = 0$$

constitute a system of homogeneous linear wave equations for the Γ_μ. Consequently the Γ_μ vanish identically provided that the initial conditions vanish, that is

$$\Gamma_\mu|_{\Sigma_0} = 0, \tag{54}$$

$$\partial_0 \Gamma_\mu|_{\Sigma_0} = 0, \tag{55}$$

where Σ_0 is the initial hypersurface $x^0 = 0$. Given now initial data for the Einstein equations

$$R_{\mu\nu} = 0 \quad \text{or, equivalently,} \quad \hat{R}_{\mu\nu} = 0, \tag{56}$$

where

$$\hat{R}_{\mu\nu} = R_{\mu\nu} - \frac{1}{2} g_{\mu\nu} R,$$

we construct initial data

$$g_{\mu\nu}|_{\Sigma_0}, \tag{57}$$

$$\partial_0 g_{\mu\nu}|_{\Sigma_0} \tag{58}$$

for the reduced equations $H_{\mu\nu} = 0$, such that the conditions

$$\Gamma_\mu|_{\Sigma_0} = 0, \quad \partial_0 \Gamma_\mu|_{\Sigma_0} = 0$$

are satisfied. Then the solution $g_{\mu\nu}$ of the Cauchy problem for the reduced equations shall, according to the above, also satisfy the conditions $\Gamma_\mu = 0$, therefore shall be a solution of the original Einstein equations.

Initial data for the Einstein equations consist of a pair $(\bar{g}_{ij}, k_{ij})$ where $\bar{g}_{ij}$ is a Riemannian metric and k_{ij} a 2-covariant symmetric tensor field on the 3-manifold $\overline{M}$, which is to be identified with the initial hypersurface Σ_0. Once we have a solution (M, g) with $M = [0, T) \times \Sigma_0$ and $\Sigma_0 = \overline{M}$, then $\bar{g}_{ij}$ and k_{ij} shall be, respectively, the 1st and 2nd fundamental form of $\Sigma_0 = \{0\} \times \Sigma_0$ in (M, g). Thus

$$\bar{g}_{ij} = g_{ij}|_{\Sigma_0}, \quad i, j = 1, 2, 3.$$

We choose the coordinates to be Gaussian normal along Σ_0, that is

$$g_{i0}|_{\Sigma_0} = 0, \tag{59}$$

$$g_{00}|_{\Sigma_0} = -1. \tag{60}$$

Then

$$\partial_0 g_{ij}|_{\Sigma_0} = 2k_{ij}. \tag{61}$$

We then choose

$$\partial_0 g_{0i}|_{\Sigma_0}, \quad \partial_0 g_{00}|_{\Sigma_0}$$

so as to satisfy the conditions

$$\Gamma_\mu|_{\Sigma_0} = 0.$$

A short calculation shows that $\partial_0 g_{0i}|_{\Sigma_0} = \overline{\Gamma}_i$ and $\partial_0 g_{00}|_{\Sigma_0} = 2\operatorname{tr} k$, where $\overline{\Gamma}_i$ are the corresponding (3-dimensional) quantities for the induced metric $\bar{g}_{ij}$. (Recall: $\operatorname{tr} k = \bar{g}^{ij} k_{ij}$.) This completes the specification of initial data for the reduced equations. Now consider the following: For a solution of the reduced equations,

$$\hat{R}_{0i}|_{\Sigma_0} = \frac{1}{2}\hat{S}_{0i}|_{\Sigma_0} = \frac{1}{2}\{\partial_0\Gamma_i + \partial_i\Gamma_0 - g_{0i}(\partial^\lambda\Gamma_\lambda\}_{|\Sigma_0} = \frac{1}{2}\partial_0\Gamma_i|_{\Sigma_0}$$

and

$$\hat{R}_{00}|_{\Sigma_0} = \frac{1}{2}\hat{S}_{00}|_{\Sigma_0} = \frac{1}{2}\{2\partial_0\Gamma_0 - g_{00}(\partial^\lambda\Gamma_\lambda\}_{|\Sigma_0} = \frac{1}{2}\partial_0\Gamma_0|_{\Sigma_0}.$$

Therefore, if the initial data $(\bar{g}_{ij}, k_{ij})$ verify the *constraint* equations

$$\hat{R}_{0i}|_{\Sigma_0} = 0, \tag{62}$$

$$\hat{R}_{00}|_{\Sigma_0} = 0, \tag{63}$$

then the conditions $\partial_0\Gamma_\mu|_{\Sigma_0} = 0$ are satisfied as well.

In the original work of Choquet-Bruhat a local problem was posed, the initial data being given on a domain $\Omega \subset \Sigma_0$. As a first step the initial data for the reduced equations is extended to the whole of $\mathbb{R}^3$ in such a way that it becomes trivial outside a larger domain Ω', where $\Omega' \supset \Omega$, with compact closure in $\mathbb{R}^3$.

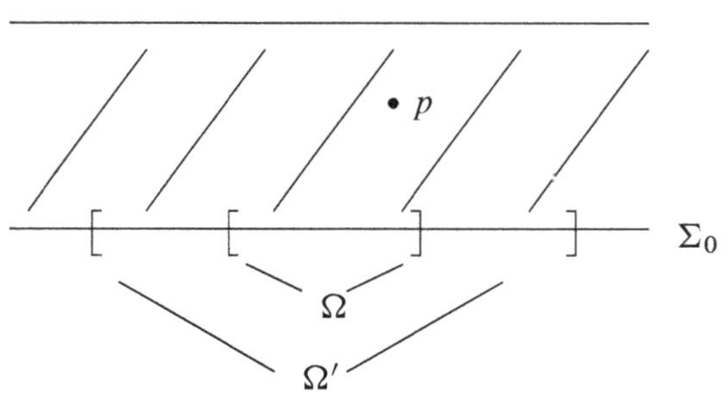

Figure 4

The next step in the construction of the solution is based on the domain of dependence theorem, to be formulated below. Let (M, g) be the known spacetime, where $M = [0, T] \times \Sigma_0$.

Definition 23. The domain of dependence of Ω in the spacetime (M, g) is the subset of M for which Ω is a (incomplete) Cauchy hypersurface.

So, the domain of dependence $\mathcal{D}(\Omega)$ of Ω in M is the set of points $p \in M$ such that each past-directed causal curve in M through p intersects Ω. (It follows that it cannot intersect more than once.)

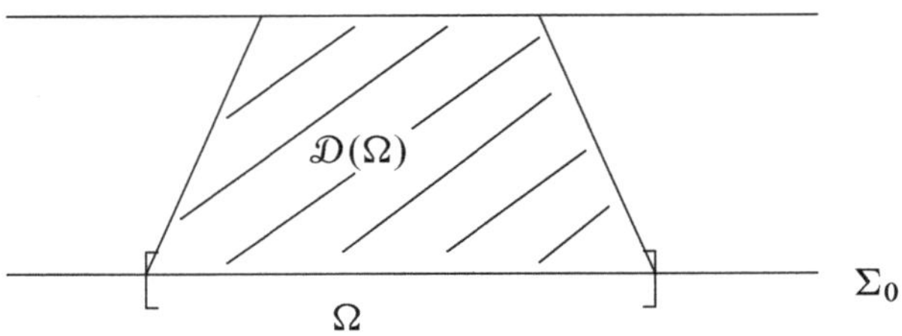

Figure 5

In $\mathcal{D}(\Omega)$ the solution depends only on the initial data in Ω. In particular, since the constraint equations are satisfied in Ω, we have that Γ_μ and $\partial_0 \Gamma_\mu$ all vanish in Ω. So, by the domain of dependence theorem applied to the (linear homogeneous) wave equations for Γ_μ, the Γ_μ vanish throughout $\mathcal{D}(\Omega)$. Therefore the solution of the reduced equations is in fact a solution of the Einstein equations

$$R_{\mu\nu} = 0 \quad \text{in } \mathcal{D}(\Omega).$$

If the 3-manifold $\overline{M}$ is compact, one can cover $\overline{M}$ with a finite number of coordinate charts and construct a local time solution by putting together these local solutions. This works by virtue of the domain of dependence theorem for the Einstein equations. For, suppose that Ω_1 and Ω_2 are two such coordinate charts with $\Omega_1 \cap \Omega_2 \neq \emptyset$. Since we are given initial data $(\bar{g}, k)$ on the whole 3-manifold $\overline{M}$, the representations of these data, given by the two charts, are related by the diffeomorphism in the overlap. Thus there exists a diffeomorphism f of $\Omega_1 \cap \Omega_2$ onto itself such that

$$\bar{g}_2 = f^* \bar{g}_1, \quad k_2 = f^* k_1.$$

After making this transformation we may assume that the initial data coincide in $\Omega_1 \cap \Omega_2$. If g_1 and g_2 are the two solutions of the reduced equations, corresponding to the initial conditions in Ω_1 and Ω_2 respectively, the domain of dependence theorem says that g_1 and g_2 coincide in the domain of dependence of $\Omega_1 \cap \Omega_2$ relative to either g_1 or g_2.

We can therefore extend either solution

$$(\mathcal{D}(\Omega_1), g_1), \quad (\mathcal{D}(\Omega_2), g_2)$$

to the union, the domain of dependence of $\Omega_1 \cup \Omega_2$.

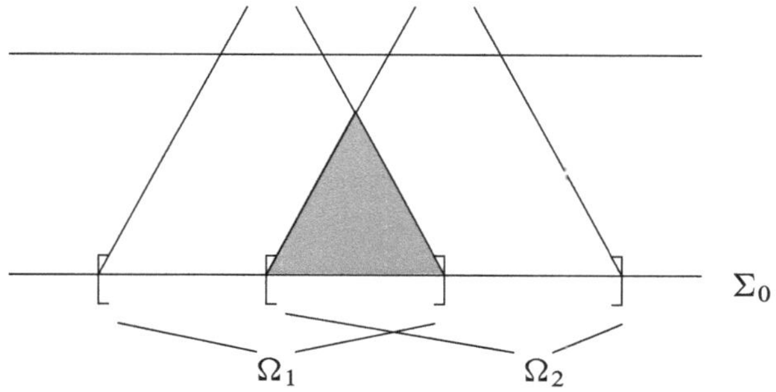

Figure 6

Remark. The domain of dependence theorem is a refined uniqueness theorem.

Definition 24. Given initial data $(\Omega, \bar{g}, k)$ where $(\Omega, \bar{g})$ is not required to be complete, we say that a spacetime $(\mathcal{U}, g)$ is a development of this data, if Ω is a Cauchy hypersurface for $\mathcal{U}$. So Ω is the past boundary of $\mathcal{U}$ and $\bar{g}$ and k are respectively the first and second fundamental forms of the hypersurface Ω in $(\mathcal{U}, g)$. Moreover, g satisfies the Einstein equations

$$R_{\mu\nu} = 0.$$

An argument analogous to the one just presented shows that if $\mathcal{U}_1$ and $\mathcal{U}_2$ are developments of the data $(\Omega, \bar{g}, k)$, then we can define a development with domain $\mathcal{U}_1 \cup \mathcal{U}_2$ which extends the corresponding metrics g_1 and g_2. Therefore, the union of all developments of given initial data is also a development of the same data, the *maximal development* of that data.

Theorem 2 (Y. Choquet-Bruhat, R. Geroch [9]). *Any initial data set $(\overline{M}, \bar{g}, k)$ (completeness of $(\overline{M}, \bar{g})$ not assumed) satisfying the constraint equations, gives rise to a unique maximal development.*

We shall now give an exposition of the domain of dependence theorem in a general Lagrangian setting. Recall the Lagrangian for a mapping $u \colon M \to \mathcal{N}$ from Section 2.1.1. Given a background solution u_0, we defined the quadratic form $h = \frac{\partial^2 L}{\partial v^2}(v_0)$ with $h(\dot{v}, \dot{v}) = h^{\mu\nu}_{ab} \dot{v}^a_\mu \dot{v}^b_\nu$ where $h^{\mu\nu}_{ab} = \frac{\partial^2 L}{\dot{v}^a_\mu \dot{v}^b_\nu}$ and $v_0 = du_0(x)$. Let us denote by $\{L\}$ the equivalence class of Lagrangians giving rise to the same Euler–Lagrange equations. Recall Definition 19 from Section 2.1.1.

Definition 25. We say that $\{L\}$ is regularly hyperbolic at v_0 if the quadratic form h fulfills the following:

1. There is a covector $\xi \in T_x^* M$ such that h is negative definite on
$$L_\xi := \{\xi \otimes Q : Q \in T_q \mathcal{N}\} \subset \mathcal{L}(T_x M, T_q \mathcal{N}),$$
that is, the elements of the form $\dot{v}^a_\mu = \xi_\mu Q^a$.
2. There is a vector $X \in T_x M$ with $\xi \cdot X > 0$ such that h is positive definite on the set Σ^1_X of rank-1-elements of the subspace
$$\Sigma_X = \{\dot{v} \in \mathcal{L}(T_x M, T_q \mathcal{N}) : \dot{v} \cdot X = 0\},$$
that is, the elements of the form $\dot{v}^a_\mu = \zeta_\mu P^a$ where $\zeta_\mu X^\mu = 0$.

Definition 26. Let h be regularly hyperbolic. Set
$$I_x^* = \{\xi \in T_x^* M : h \text{ is negative definite on } L_\xi\} \tag{64}$$
and
$$J_x = \{X \in T_x M : h \text{ is positive definite on } \Sigma^1_X\}. \tag{65}$$

Proposition 2. *I_x^* and J_x are open cones each of which has two components*
$$I_x^* = I_x^{*+} \cup I_x^{*-},$$
$$J_x = J_x^+ \cup J_x^-,$$
where I_x^{*-} and J_x^- are the sets of opposites of elements in I_x^{*+} and J_x^+, respectively. Moreover, each component is convex. The boundary ∂I_x^* is a component (the inner component) of C_x^*, the characteristic in $T_x^* M$, and ∂J_x is a component (the inner component) of C_x, the characteristic in $T_x M$.

Proof. The proof is in the book [12]. □

Recall the definition of the Noether transform $m(\xi, X)$ of h corresponding to a pair $(\xi, X) \in T_x^* M \times T_x M$ with $\xi \cdot X > 0$:
$$m(\xi, X)(\dot{v}_1, \dot{v}_2) = (\xi, X) h(\dot{v}_1, \dot{v}_2) - h(\xi \otimes \dot{v}_1 \cdot X, \dot{v}_2) - h(\dot{v}_1, \xi \otimes \dot{v}_2 \cdot X).$$

Proposition 3. *Let $\mathcal{U}_x^+ \subset T_x^* M \times T_x M$ be given by*
$$\mathcal{U}_x^+ = \{(\xi, X) : \xi \cdot X > 0\}.$$
Consider the subset of $\mathcal{U}_x^+$ consisting of those (ξ, X) with $m(\xi, X)$ positive definite on $R_\xi = \{\xi \otimes P + \zeta \otimes Q : \forall\, \zeta \in T_x^* M,\ \forall\ P, Q \in T_q \mathcal{N}\}$. Then this subset is given by
$$(I_x^{*+} \times J_x^+) \cup (I_x^{*-} \times J_x^-).$$
Moreover, on the boundary of this set $m(\xi, X)$ has nullity.

Let us discuss briefly the notion of variation of a mapping $u_0 \colon M \to \mathcal{N}$. A variation of u_0, namely $\dot{u}$, is a section of $u_0^* T\mathcal{N}$ (the pullback by u_0 of $T\mathcal{N}$). In general, if $\mathcal{B}$ is a bundle over $\mathcal{N}$ and $u_0 \colon M \to \mathcal{N}$, we denote by $u_0^*\mathcal{B}$ the *pullback bundle*, namely the following bundle over M:

$$u_0^*\mathcal{B} = \bigcup_{x \in M} \{x\} \times \mathcal{B}_{u_0(x)},$$

where $\mathcal{B}_q$ is the fibre of $\mathcal{B}$ over $q \in \mathcal{N}$.

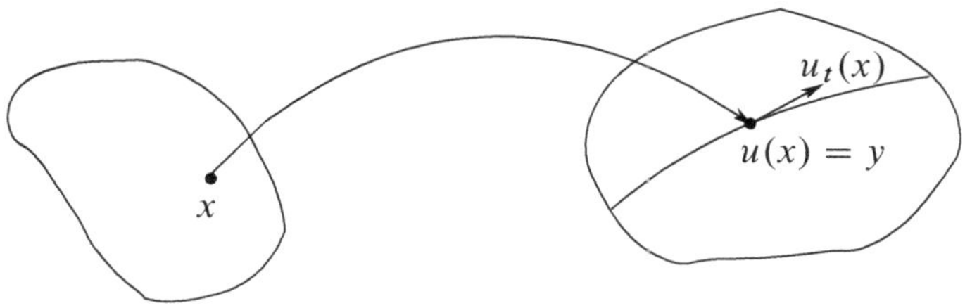

Figure 7

Thus, a variation $\dot{u}$ maps $x \in M \mapsto \dot{u}(x) \in T_{u_0(x)}\mathcal{N}$. For a given $x \in M$, $\dot{u}(x)$ is the tangent vector at $u_0(x)$ of the curve $t \mapsto u_t(x)$ in $\mathcal{N}$, where u_t is a differentiable 1-parameter family of mappings $u_t \colon M \to \mathcal{N}$, i.e., $\dot{u}(x) = \frac{du_t(x)}{dt}\big|_{t=0}$.

We now explain the meaning of the subset $J_x \subset T_x M$: J_x is the set of possible values at $x \in M$ of a vector field X on M with the property that the reduced equations obtained by considering mappings which are invariant under the corresponding 1-parameter group of diffeomorphisms of M, form a regularly elliptic system.

On the other hand, the subset $I_x^* \subset T_x^* M$ defines the notion of a spacelike hypersurface.

Definition 27. A hypersurface H in M is called spacelike, if at each $x \in H$ the double ray $\{\lambda\xi : \lambda \neq 0 \in \mathbb{R}\}$ defined by the hyperplane $T_x H$ in $T_x M$,

$$T_x H = \{Y \in T_x M : \xi \cdot Y = 0\}, \tag{66}$$

is contained in I_x^*.

Definition 28. $\bar{I}_x$, the causal subset of $T_x M$, is the set of all vectors $X \in T_x M$ such that $\xi \cdot X \neq 0$ for all $\xi \in I_x^*$.

$\bar{I}_x$ is a closed subset of $T_x M$ with $\bar{I}_x = \bar{I}_x^+ \cup \bar{I}_x^-$, where $\bar{I}_x^-$ is the set of opposites of elements in $\bar{I}_x^+$. One can show that each component is convex. If $X \in \partial \bar{I}_x$ then there exists a covector $\xi \in \partial I_x^*$ such that $\xi \cdot X = 0$. It follows that each component I_x^{*+} and I_x^{*-} lies to one side of the plane

$$\Pi_X = \{\xi \in T_x^* M : \xi \cdot X = 0\}$$

and Π_X contains a ray of ∂I_x^{*+} respectively ∂I_x^{*-}. Thus if ∂I_x^* is differentiable at this double ray, then Π_X is the tangent plane to I_x^* at this double ray.

Definition 29. A causal curve γ in M is a curve in M whose tangent vector $\dot{\gamma}(t)$ at each point $\gamma(t)$ belongs to $\bar{I}_{\gamma(t)}$.

The following statements are valid for the future and past components of $\bar{I}_x$ and J_x separately. In general, we have $\bar{I}_x \supset J_x$. In fact, J_x is the interior of the inner component of C_x while $\bar{I}_x$ is the convex hull of the outer component of C_x. (Remember that C_x is the characteristic in the tangent space $T_x M$.)

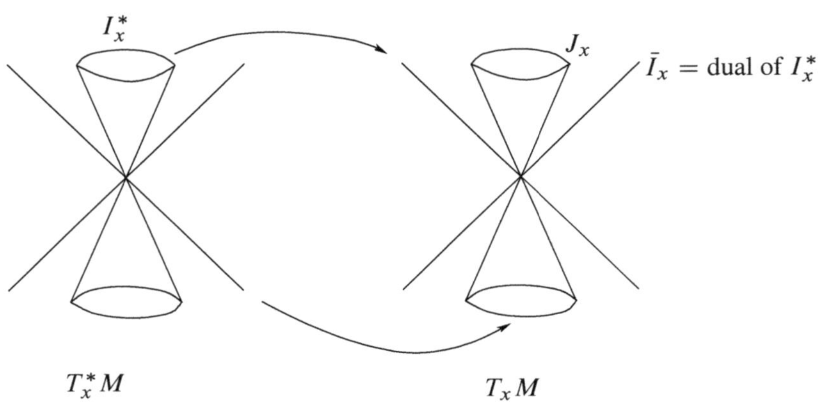

Figure 8

We have in general m-sheets in the case dim $\mathcal{N} = m$.

Next, let us give two equivalent definitions of *domain of dependence*:

Definition 30. Let $\mathcal{R}$ be a domain in M on which a solution u of the Euler–Lagrange equations is defined. Consider a domain $\mathcal{D} \subset \mathcal{R}$ and a hypersurface Σ in $\mathcal{R}$, which is spacelike relative to du.

1. We say that $\mathcal{D}$ is a development of Σ if we can express

$$\mathcal{D} = \bigcup_{t \in [0,T]} \Sigma_t$$

where $\{\Sigma_t : t \in [0, T]\}$ is a foliation and where each Σ_t is a spacelike (relative to du) hypersurface in $\mathcal{R}$ homologous to $\Sigma_0 = \Sigma$. (In particular, $\partial\Sigma_t = \partial\Sigma$ for all $t \in [0, T]$.)

2. $\mathcal{D}$ is a development of Σ if each causal curve in $\mathcal{R}$ through any point of $\mathcal{D}$ intersects Σ at a single point.

We sketch the proof of the equivalence of the above two definitions.

Sketch of proof. Suppose that $\mathcal{D}$ is a development of Σ according to Definition 30.2. Then for each point $p \in \mathcal{D}$ the causal past $J^-(p)$ (we are considering future developments; Σ is the past boundary of $\mathcal{D}$) is compact. We define $t(p)$ to be the volume of $J^-(p)$, given a volume form ϵ on $\mathcal{R}$. This defines a time function in $\mathcal{D} \setminus \Sigma$: $dt \cdot X > 0$ for any vector $X \in \bar{I}_x^+$ with $x \in \mathcal{D} \setminus \Sigma$. The level sets of t then define a foliation $\{\Sigma_t\}$ as required in Definition 30.1. Conversely, suppose that $\mathcal{D}$ is a development of Σ according to Definition 30.1. Then any causal curve in $\mathcal{D}$ can be parametrized by t in a non-singular manner. It follows that each past-directed causal curve from any point of $\mathcal{D}$ must intersect Σ. One can show that if $\mathcal{D}_1$ and $\mathcal{D}_2$ are developments of Σ, then $\mathcal{D}_1 \cup \mathcal{D}_2$ is also a development. So given a domain of definition $\mathcal{R}$ and a spacelike hypersurface Σ, we can define the *domain of dependence* of Σ in $\mathcal{R}$ relative to du to be the *maximal development.* □

To state the domain of dependence theorem in a precise and general manner, we reformulate the general Lagrangian setup from a global perspective. Consider maps $u\colon M \to \mathcal{N}$. The *configuration space* is $C = M \times \mathcal{N}$. The *velocity space* $\mathcal{V}$ is a bundle over C:

$$\mathcal{V} = \bigcup_{(x,q) \in C} \mathcal{L}(T_x M, T_q \mathcal{N}).$$

We have a projection

$$\pi : \mathcal{V} \to C,$$
$$v \in \mathcal{L}(T_x M, T_q \mathcal{N}) \mapsto (x, q) \in C,$$

where we write $\pi_{\mathcal{V},M}$ for $\pi_1 \circ \pi : \mathcal{V} \to C \to M$ and we write $\pi_{\mathcal{V},\mathcal{N}}$ for $\pi_2 \circ \pi : \mathcal{V} \to C \to \mathcal{N}$.

Let us list at this point the relevant notions.

TM,	the tangent bundle of M,
$\Lambda_r M$,	the bundle of (fully antisymmetric) r-forms on M,
$S_2 M$,	the bundle of quadratic (symmetric bilinear) forms on M,

and, with $n = \dim M$,

$\Lambda_n M$,	the bundle of top-degree-forms on M.

So, $\Lambda_n M$ is a bundle over M and

$$\pi_{\mathcal{V},M} : \mathcal{V} \to M$$

is the projection defined above. Then the pullback bundle is

$$\pi^*_{\mathcal{V},M}\ \Lambda_n M,$$

a bundle over $\mathcal{V}$. An element of this is $\omega \in (\Lambda_n M)_x$ with $x \in M$, and this ω is attached to an element $v \in \mathcal{L}(T_x M, T_q \mathcal{N})$. The Lagrangian L is actually a C^∞-section of $\pi^*_{\mathcal{V},M}\ \Lambda_n M$ over $\mathcal{V}$. So,

$$v \mapsto L(v) \in (\Lambda_n M)_x; \quad v \in \mathcal{L}(T_x M, T_q \mathcal{N}).$$

Thus

$$L(v)(Y_1, \dots, Y_n),$$

with $Y_1, \dots, Y_n \in T_x M$, is an n-linear fully antisymmetric form on $T_x M$.

Definition 31. Given a Lagrangian L, the action of a map u defined in a domain $\mathcal{R}$ in M, corresponding to a subdomain $\mathcal{D} \subset \mathcal{R}$ is

$$\mathcal{A}\,[u; \mathcal{D}] = \int_{\mathcal{D}} L \circ du. \tag{67}$$

Note that $L \circ du$ is a section of $\Lambda_n M$ over $\mathcal{R}$ ($(L \circ du)(x) = L(du(x)) \in (\Lambda_n M)_x$). Also, note that with $\dim M = n$, $\dim \mathcal{N} = m$, we have $\dim C = n + m$, $\dim \mathcal{V} = (n + m) + nm$.

Suppose that M is oriented and ϵ is a C^∞ volume form on M. That is, ϵ is a C^∞-section of $\Lambda_n M$ such that if $(E_1, \dots, E_n)$ is a positive basis for $T_x M$ (a positive frame at x), then $\epsilon(E_1, \dots, E_n) > 0$. Given then a C^∞ function L^* on $\mathcal{V}$, we define the corresponding Lagrangian L by

$$L(v)(Y_1, \dots, Y_n) = L^*(v)\, \epsilon(Y_1, \dots, Y_n)$$

with $v \in \mathcal{L}(T_x M, T_q \mathcal{N})$ and $Y_1, \dots, Y_n$ in $T_x M$. The function L^* was called "Lagrangian" in the previous.

Finally, we state the domain of dependence theorem.

Theorem 3 (Domain of dependence theorem). *Let u_0 be a C^2 solution of the Euler–Lagrange equations corresponding to a C^∞ Lagrangian L; and let u_0 be defined in a domain $\mathcal{R}$ in M. Let Σ be a hypersurface in $\mathcal{R}$, which is spacelike relative to du_0. Let also u_1 be another solution of the Euler–Lagrange equations defined and C^1 on $\mathcal{R}$. Suppose that*

$$du_0|_\Sigma = du_1|_\Sigma.$$

Then u_1 coincides with u_0 in the domain of dependence of Σ in $\mathcal{R}$ relative to du_0.

2.3 Decomposition of the Einstein equations with respect to the foliation by the level sets $\mathcal{H}_t$ of a time function t

We consider first Lorenzian geometry without the Einstein equations, and afterwards we impose the Einstein equations.

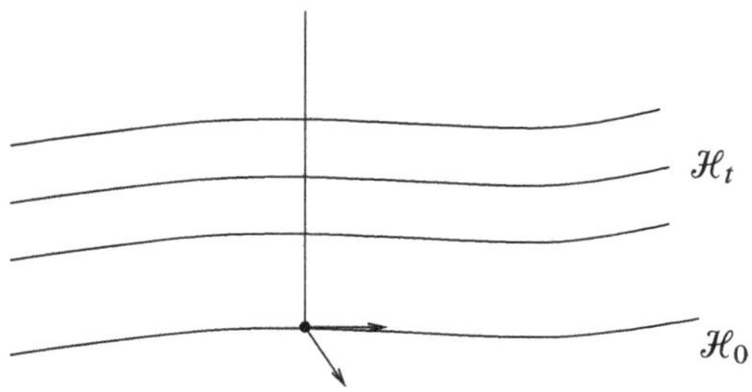

Figure 9

Let (E_1, E_2, E_3) be a (local) frame field for $\mathcal{H}_0$. We extend the E_i, $i = 1, 2, 3$, to the spacetime by the condition

$$[T, E_i] = 0. \tag{68}$$

Then the (E_1, E_2, E_3) define a frame field for each $\mathcal{H}_t$. The spacetime manifold M is represented as the product $[0, T] \times \mathcal{H}_0$. In this representation $T = \frac{\partial}{\partial t}$ and

$$g = -\Phi^2 dt^2 + \bar{g},$$

where $\bar{g}(t)$ is the induced metric on $\mathcal{H}_t$ and Φ is the lapse function. Recall the first variation equations

$$\frac{\partial \bar{g}_{ij}}{\partial t} = 2\Phi k_{ij}, \quad k_{ij} = k(E_i, E_j). \tag{69}$$

Here k is the second fundamental form of $\mathcal{H}_t$ and the indices i, j refer to the frame $(E_i : i = 1, 2, 3)$. The second variation equations are

$$\frac{\partial k_{ij}}{\partial t} = \bar{\nabla}_i \bar{\nabla}_j \Phi - (R_{i0j0} - k_{im} k_j^m)\Phi. \tag{70}$$

Here $\bar{\nabla}$ is the covariant derivative operator intrinsic to $\mathcal{H}_t$, that is, relative to the Riemannian connection of $(\mathcal{H}_t, \bar{g}(t))$. Also, $R_{i0j0} = R(E_i, E_0, E_j, E_0)$, the frame field (E_1, E_2, E_3) being completed by $E_0 = \frac{1}{\Phi}\frac{\partial}{\partial t}$, the future-directed unit normal to the $\mathcal{H}_t$, to a frame field for M. Remark that $g_{00} = -1$, $g_{0i} = 0$ and $g_{ij} = \bar{g}_{ij} = g(E_i, E_j)$.

In addition to the 1st and 2nd variation equations we have the Codazzi and Gauss equations of the embedding of $\mathcal{H}_t$ in M. The Codazzi equations are

$$\overline{\nabla}_i k_{jm} - \overline{\nabla}_j k_{im} = R_{m0ij}, \tag{71}$$

and the Gauss equations are

$$\overline{R}_{imjn} + k_{ij}k_{mn} - k_{in}k_{mj} = R_{imjn}. \tag{72}$$

Here $\overline{R}_{imjn}$ are the components of the curvature tensor of $(\mathcal{H}_t, \bar{g}(t))$.

We proceed to impose the Einstein equations. Taking the trace of the Gauss equations we obtain

$$\overline{R}_{ij} + \operatorname{tr} k\, k_{ij} - k_{im}k^m_j = R_{ij} + R_{i0j0}. \tag{73}$$

Next we substitute for R_{i0j0} from these equations into the 2nd variation equations to conclude that the part $R_{ij} = 0$ of the vacuum Einstein equations is equivalent to

$$\frac{\partial k_{ij}}{\partial t} = \overline{\nabla}_i \overline{\nabla}_j \Phi - (\overline{R}_{ij} + k_{ij} \operatorname{tr} k - 2k_{im}k^m_j)\Phi. \tag{74}$$

The trace of the Codazzi equations is

$$\overline{\nabla}^i k_{ij} - \partial_j \operatorname{tr} k = R_{0j}. \tag{75}$$

The part $R_{0j} = 0$ of the vacuum Einstein equations is thus equivalent to the constraint equation

$$\overline{\nabla}^i k_{ij} - \partial_j \operatorname{tr} k = 0. \tag{76}$$

The double trace of the Gauss equations is

$$\overline{R} + (\operatorname{tr} k)^2 - |k|^2 = R + 2R_{00} = 2\hat{R}_{00}, \tag{77}$$

where $|k|^2 = k^i_m k^m_i$ and $\hat{R}_{\mu\nu} = R_{\mu\nu} - \frac{1}{2} g_{\mu\nu} R$. Thus the vacuum Einstein equation $\hat{R}_{00} = 0$ is equivalent to the constraint equation

$$\overline{R} + (\operatorname{tr} k)^2 - |k|^2 = 0. \tag{78}$$

The constraint equations constrain the initial data

$$(\bar{g}, k) \quad \text{on } \mathcal{H}_0.$$

To derive the 2nd variation equations we must obtain an expression for the acceleration of the orthogonal family of curves.

Acceleration of the integral curves of T. Let us denote the unit normal $E_0 = N$. Then the geodesic curvature of the orthogonal family of curves is given by

$$\nabla_N N = \Phi^{-1}\overline{\nabla}\Phi, \tag{79}$$

and $\overline{\nabla}\Phi$ is the gradient of Φ intrinsic to $\mathcal{H}_t$, a vector field tangent to $\mathcal{H}_t$ with components

$$\overline{\nabla}^i\Phi = \bar{g}^{ij}\overline{\nabla}_j\Phi,$$

where $\overline{\nabla}_j\Phi = E_j\Phi$. Recall that $T = \Phi N$ (the integral curves of T and N are the same, but differ only by parametrization; T is parametrized by t while N is parametrized by s, namely arc length.) The formula (79) is derived as follows. We have (in arbitrary local coordinates)

$$N_\mu = g_{\mu\nu}N^\nu = -\Phi\partial_\mu t. \tag{80}$$

Hence, we can write

$$\begin{aligned} N^\nu\nabla_\nu N_\mu &= -N^\nu\nabla_\nu(\Phi\partial_\mu t) \\ &= -N^\nu\partial_\nu\Phi\partial_\mu t - \Phi N^\nu\nabla_\mu(\partial_\nu t) \\ &= \Phi^{-1}N^\nu N_\mu\partial_\nu\Phi + N^\nu\Phi\nabla_\mu(\Phi^{-1}N_\nu). \end{aligned}$$

The last term is

$$-\Phi^{-1}\underbrace{N^\nu N_\nu}_{=-1}\partial_\mu\Phi + \underbrace{N^\nu\nabla_\mu N_\nu}_{=\frac{1}{2}\partial_\mu\underbrace{(N^\nu N_\nu)}_{=-1}=0} = \Phi^{-1}\partial_\mu\Phi.$$

Thus,

$$N^\nu\nabla_\nu N^\mu = \Phi^{-1}\Pi^{\mu\nu}\partial_\nu\Phi,$$

where $\Pi^{\mu\nu} = g^{\mu\nu} + N^\mu N^\nu = g^{\nu\lambda}\Pi^\mu_\lambda$ and Π^μ_λ defines the orthogonal projection to the $\mathcal{H}_t$:

$$\Pi\cdot X = X + g(N,X)N$$

on any vector $X \in TM$. We have thus obtained the formula (79). Since $T = \Phi N$, an expression for the acceleration $\nabla_T T$ of the integral curves of T readily follows.

Now let X, Y be vector fields tangential to the $\mathcal{H}_t$ and satisfying

$$[T,X] = [T,Y] = 0.$$

At the end we shall set $X = E_i$ and $Y = E_j$ to obtain $Tk(E_i,E_j) = \frac{\partial k_{ij}}{\partial t}$. We have

$$\begin{aligned} k(X,Y) &= g(\nabla_X N, Y) \\ &= \Phi^{-1}g(\nabla_X T, Y). \end{aligned}$$

Therefore

$$T(k(X,Y)) = T(\Phi^{-1} g(\nabla_X T, Y)).$$

Now,

$$T(g(\nabla_X T, Y)) = g(\nabla_T \nabla_X T, Y) + g(\nabla_X T, \nabla_T Y).$$

For the first term we apply the definition of the curvature:

$$\nabla_T \nabla_X T = \nabla_X \nabla_T T + \nabla_{[T,X]} T + R(T,X)T$$

and

$$g(Y, R(T,X)T) = R(Y,T,T,X) = -R(X,T,Y,T).$$

For the second term we use

$$\nabla_T Y - \nabla_Y T = [T,Y] = 0,$$

to write

$$g(\nabla_X T, \nabla_T Y) = g(\nabla_X T, \nabla_Y T).$$

Substituting $T = \Phi N$, we can express the vector field $Z := \nabla_X N$ in terms of k. In fact Z is tangential to the $\mathcal{H}_t$, hence $Z = Z^i E_i$ and $Z^i g_{ij} = g(Z, E_j) = k(X, E_j)$. The 2nd variation equations (70) follow in this manner after substituting for $\nabla_T T$ from the formula (79) for $\nabla_N N$.

To derive the Gauss equations, we recall from Riemannian geometry that the covariant derivative $\overline{\nabla}$ intrinsic to $(\mathcal{H}_t, \bar{g}(t))$ is characterized by the property

$$\overline{\nabla}_X Y = \Pi \cdot \nabla_X Y,$$

where X, Y are any vector fields tangential to the $\mathcal{H}_t$. We apply this to the definition of the curvature of $(\mathcal{H}_t, \bar{g}(t))$. Let X, Y, Z be arbitrary vector fields tangential to $\mathcal{H}_t$. Then

$$\overline{R}(X,Y)Z = \overline{\nabla}_X \overline{\nabla}_Y Z - \overline{\nabla}_Y \overline{\nabla}_X Z - \overline{\nabla}_{[X,Y]} Z. \tag{81}$$

Now let W be another arbitrary vector field tangential to $\mathcal{H}_t$. Then we can write

$$\overline{R}(W,Z,X,Y) = g(W,\ \overline{R}(X,Y)Z) \tag{82}$$

$$= g(W,\ \nabla_X \overline{\nabla}_Y Z - \nabla_Y \overline{\nabla}_X Z - \nabla_{[X,Y]} Z), \tag{83}$$

since $g(W, \Pi \cdot U) = g(W, U)$ for any vector field U. At this point we substitute

$$\overline{\nabla}_X Z = \Pi \cdot \nabla_X Z = \nabla_X Z + g(N, \nabla_X Z) N.$$

Thus in $\nabla_Y(\overline{\nabla}_X Z)$ there are terms, in addition to $\nabla_Y \nabla_X Z$, which involve $\nabla_Y N$, thus the second fundamental form k. The Gauss equations (72) follow in this manner. The Codazzi equations (71) are straightforward.

3 Asymptotic flatness at spacelike infinity and conserved quantities in General Relativity

3.1 Conserved quantities

In this chapter we discuss the definitions of total energy, linear momentum and angular momentum in General Relativity. We then give an overview followed by a rigorous discussion of the associated conservation laws.

We begin with the definitions of a manifold which is Euclidean at infinity, of an asymptotically Euclidean Riemannian manifold and of an asymptotically flat initial data set.

Definition 32. A 3-manifold $\mathcal{H}$ is said to be Euclidean at infinity if there exists a compact set $\mathcal{K} \subset \mathcal{H}$ such that $\mathcal{H} \setminus \mathcal{K}$ is diffeomorphic to $\mathbb{R}^3 \setminus \mathcal{B}$, where $\mathcal{B}$ is a ball in $\mathbb{R}^3$. Thus $\mathcal{H} \setminus \mathcal{K}$ is contained in the domain of a chart.

Definition 33. An asymptotically Euclidean Riemannian manifold $(\mathcal{H}, \bar{g})$ is a complete Riemannian manifold which is Euclidean at infinity and there exists a coordinate system (x^1, x^2, x^3) in the complement of $\mathcal{K}$ above relative to which the metric components $\bar{g}_{ij} \to \delta_{ij}$ as $r := \sqrt{\sum_{i=1}^{3}(x^i)^2} \to \infty$.

Definition 34. An asymptotically flat initial data set $(\mathcal{H}, \bar{g}, k)$ is an initial data set where $(\mathcal{H}, \bar{g})$ is an asymptotically Euclidean Riemannian manifold and the components of k approach 0 relative to the coordinate system above as $r \to \infty$.

The fall-off of $\bar{g}_{ij} - \delta_{ij}$ and k_{ij} with r should be sufficiently rapid for the notions of total energy, linear momentum and angular momentum below to be well defined and finite.

The standard definition of these notions is the following:

Definition 35 (Arnowitt, Deser, Misner (ADM) [1]). Let $S_r = \{|x| = r\}$ be the coordinate sphere of radius r and dS_j the Euclidean oriented area element of S_r. We then define

- the total energy

$$E = \frac{1}{4} \lim_{r \to \infty} \int_{S_r} \sum_{i,j} (\partial_i \bar{g}_{ij} - \partial_j \bar{g}_{ii})\, dS_j, \tag{84}$$

- the linear momentum

$$P^i = -\frac{1}{2}\lim_{r\to\infty}\int_{S_r}(k_{ij} - \bar{g}_{ij}\,\mathrm{tr}\,k)\,dS_j, \tag{85}$$

- the angular momentum

$$J^i = -\frac{1}{2}\lim_{r\to\infty}\int_{S_r}\epsilon_{ijm}x^j(k_{mn} - \bar{g}_{mn}\,\mathrm{tr}\,k)\,dS_n. \tag{86}$$

The total energy, linear momentum and angular momentum are conserved quantities at spacelike infinity, as shall be shown in the sequel.

We now begin the discussion of where these quantities come from. Let us recall the fundamental theorem of Noether.

Theorem 4 (Noether's theorem [22]). *In the framework of a Lagrangian theory, to each continuous group of transformations leaving the Lagrangian invariant there corresponds a quantity which is conserved.*

In particular:

- energy corresponds to time translations,
- linear momentum corresponds to space translations,
- angular momentum corresponds to space rotations.

As discussed in the previous chapter, a Lagrangian corresponds in a local coordinate description to a function

$$L^* = L^*(x, q, v).$$

We denote by x^μ (with $\mu = 1, \dots, n$) the independent variables, $q^a = u^a(x)$ (with $a = 1, \dots, m$) the dependent variables, and $v^a_\mu = \frac{\partial u^a}{\partial x^\mu}(x)$ the first derivatives of the dependent variables. The *canonical momentum* is given by

$$p^{*\mu}_a = \frac{\partial L^*}{\partial v^a_\mu}. \tag{87}$$

We define the *canonical stress* by

$$T^{*\mu}_\nu = p^{*\mu}_a v^a_\nu - L^*\delta^\mu_\nu. \tag{88}$$

In the following we restrict ourselves to transformations acting only on the domain of the independent variables. Let X^μ be a vector field generating a 1-parameter group of transformations of this domain leaving invariant the Lagrangian form

$$L^*\,d^n x, \quad d^n x = dx^1 \wedge \dots \wedge dx^n. \tag{89}$$

Then the *Noether current*

$$J^{*\mu} = T^{*\mu}_{\nu} X^{\nu} \tag{90}$$

is divergence-free, that is

$$\partial_\mu J^{*\mu} = 0. \tag{91}$$

By the divergence theorem we conclude that the following *conservation law* holds: If Σ_1 and Σ_2 are homologous hypersurfaces (in particular $\partial\Sigma_1 = \partial\Sigma_2$), then

$$\int_{\Sigma_2} J^{*\mu}\, d\Sigma_\mu = \int_{\Sigma_1} J^{*\mu}\, d\Sigma_\mu. \tag{92}$$

We now turn to General Relativity. The Einstein equations are derived from

Hilbert's action principle. This corresponds to the following Lagrangian for gravity:

$$L = -\frac{1}{4} R\, d\mu_g,$$

where R is the scalar curvature of the metric g and $d\mu_g$ the volume form of g. This Lagrangian is the only one (up to an additive constant multiple of the volume form) which gives rise to second order (Euler–Lagrange) equations. Here the metric components $g_{\mu\nu}$ are the unknown functions and R depends on their second derivatives. Moreover, it is the only geometric invariant (up to an additive constant) which contains the second derivatives only linearly. (All other invariants give rise to 4th order equations.) But Noether's theorem depends on having a Lagrangian containing only the first derivatives of the unknown functions. Thus it cannot be applied directly to Hilbert's principle.

To deal with this difficulty, Einstein and Weyl introduced the Lagrangian

$$-\frac{1}{4} R + \partial_\alpha I^\alpha = L^*,$$

where $I^\alpha = -\frac{1}{4}\sqrt{-g}\,(g^{\mu\nu}\Gamma^\alpha_{\mu\nu} - g^{\mu\alpha}\Gamma^\nu_{\mu\nu})$. Then

$$L^* = -\frac{1}{4}\sqrt{-g}\, g^{\mu\nu}(\Gamma^\beta_{\mu\alpha}\Gamma^\alpha_{\nu\beta} - \Gamma^\beta_{\mu\nu}\Gamma^\alpha_{\beta\alpha})$$

differs from Hilbert's invariant Lagrangian $-\frac{1}{4}R\sqrt{-g}$ by a divergence, therefore gives rise to the same field equations. So, one can define

$$p^{*\mu\alpha\beta} = \frac{\partial L^*}{\partial v_{\mu\alpha\beta}}$$

to be the canonical momentum, where $v_{\mu\alpha\beta} = (\partial_\mu g_{\alpha\beta})(x)$ (q^a corresponding to $g_{\alpha\beta}(x)$). The canonical stress is then

$$T^{*\mu}_{\nu} = p^{*\mu\alpha\beta} v_{\nu\alpha\beta} - L^* \delta^\mu_\nu.$$

This is called the *Einstein pseudo-tensor.* The Lagrangian L^* is invariant under translations: $x^\mu \mapsto x^\mu + c^\mu$, where the c^μ are constants. Let

$$J^{*\mu} = T_\nu^{*\mu} X^\nu, \quad \text{where } X^\nu = c^\nu \ (X = c^\nu \frac{\partial}{\partial x^\nu}),$$

so X is a vector field generating a 1-parameter group of translations. By Noether's theorem the current $J^{*\mu}$ is divergence free, that is

$$\partial_\mu J^{*\mu} = 0.$$

Since the c^μ are arbitrary constants, the following differential conservation laws hold:

$$\partial_\mu T_\nu^{*\mu} = 0.$$

Hermann Weyl wrote the following comments about these conservation laws. We quote from the book [29], p. 273:

In the original German:

> Dennoch scheint es physikalisch sinnlos zu sein, die $T_\nu^{*\mu}$ als Energiekomponenten des Gravitationsfeldes einzuführen; denn diese Grössen *bilden weder einen Tensor noch sind sie symmetrisch.* In der Tat können durch geeignete Wahl eines Koordinatensystems alle $T_\nu^{*\mu}$ an einer Stelle stets zum Verschwinden gebracht werden; man braucht dazu das Koordinatensystem nur als ein geodätisches zu wählen. Und auf der andern Seite bekommt man in einer 'Euklidischen', völlig gravitationslosen Welt bei Benutzung eines krummlinigen Koordinatensystems $T_\nu^{*\mu}$, die verschieden von 0 sind, wo doch von der Existenz einer Gravitationsenergie nicht wohl die Rede sein kann. Sind daher auch die Differentialrelationen (oben) ohne wirkliche physikalische Bedeutung, so entsteht doch aus ihnen durch *Integration über ein isoliertes System* ein invarianter Erhaltungssatz.

In English translation:

> Nevertheless it seems to be physically meaningless to introduce the $T_\nu^{*\mu}$ as energy components of the gravitational field; for, these quantities are *neither a tensor nor are they symmetric.* In fact by choosing an appropriate coordinate system all the $T_\nu^{*\mu}$ can be made to vanish at any given point; for this purpose one only needs to choose a geodesic (normal) coordinate system. And on the other hand one gets $T_\nu^{*\mu} \neq 0$ in a 'Euclidean' completely gravitationless world when using a curved coordinate system, but where no gravitational energy exists. Although the differential relations ($\partial_\mu T_\nu^{*\mu} = 0$, above) are without a physical meaning, nevertheless by *integrating them over an isolated system* one gets invariant conserved quantities.

Consider the transformation

$$\begin{aligned} x &\mapsto \bar{x} = f(x), \\ g &\mapsto f^* g. \end{aligned}$$

In components,

$$\bar{x}^\mu = f^\mu(x)$$

and

$$g_{\mu\nu} \mapsto \bar{g}_{\mu\nu}, \quad \bar{g}_{\mu\nu}(x) = \frac{\partial f^\alpha}{\partial x^\mu}\frac{\partial f^\beta}{\partial x^\nu} g_{\alpha\beta}(\bar{x})$$

($x = f^{-1}(\bar{x})$). As an example we have, in particular, linear transformations

$$\bar{x}^\mu = a^\mu_\nu x^\nu + b^\mu,$$

where a and b are constants. Then we have

$$\bar{g}_{\mu\nu}(x) = a^\alpha_\mu a^\beta_\nu g_{\alpha\beta}(\bar{x}).$$

If $\{f_t\}$ is a 1-parameter group of transformations generated by a vector field X, then

$$\frac{d}{dt} f_t^* g\Big|_{t=0} = \mathcal{L}_X g \tag{93}$$

is the Lie derivative of g with respect to X. We have

$$(\mathcal{L}_X g)_{\mu\nu} = \nabla_\mu X_\nu + \nabla_\nu X_\mu,$$

where $X_\mu = g_{\mu\nu} X^\nu$. We say that an integrated quantity Q is gauge invariant if for every such 1-parameter group $\{f_t\}$,

$$Q[f_t^* g] = Q[g]. \tag{94}$$

This requirement implies

$$\dot{Q} = \frac{d}{dt} Q[f_t^* g]\Big|_{t=0} = 0 \tag{95}$$

and conversely. In view of (93), the last reads

$$\dot{Q} := D_g Q \cdot \mathcal{L}_X g = 0. \tag{96}$$

Here, we think of Q as a differentiable function of g.

Now, the metric g is not a mapping of the (spacetime) manifold M into another manifold $\mathcal{N}$, but rather a section of a tensor bundle over M. Thus, Noether's theorem must be extended to sections of tensor bundles. In any case, we have derived, by considering translations,

$$\partial_\mu\, T_\nu^{*\mu} = 0. \tag{97}$$

Let us now consider rotations. The vector fields

$$X_{(ij)} = x^i \frac{\partial}{\partial x^j} - x^j \frac{\partial}{\partial x^i}, \quad i, j = 1, 2, 3,$$

are the generators of rotations in the (i, j)-plane. In 3-space-dimensions the vector fields

$$X_{(i)} = \epsilon_{ij\kappa} x^j \frac{\partial}{\partial x^\kappa}$$

generate rotations about the ith coordinate axis. In Minkowski spacetime the vector field

$$X_{(\alpha\beta)} = x_\alpha \frac{\partial}{\partial x^\beta} - x_\beta \frac{\partial}{\partial x^\alpha}, \quad x_\mu = \eta_{\mu\nu} x^\nu,$$

generates spacetime-rotations in the (x^α, x^β)-coordinate plane. Here $\eta = \operatorname{diag}(-1, 1, 1, 1)$ is the Minkowski metric in rectangular coordinates. We denote the vector field generating translations along the α-coordinate axis by

$$X_{(\alpha)} = \frac{\partial}{\partial x^\alpha}.$$

Now, Noether's theorem in the case of translations applies as it stands if we take the naive point of view of considering g as a matrix-valued function on $M = \mathbb{R}^4$. This yields

$$\partial_\mu J^{*\mu}_{(\alpha)} = 0, \quad \text{where } J^{*\mu}_{(\alpha)} = T_\nu^{*\mu} X^\nu_{(\alpha)} = T_\alpha^{*\mu}.$$

On the other hand, for spacetime rotations in the (α, β)-plane we have

$$J^{*\mu}_{(\alpha\beta)} = T_\nu^{*\mu} X^\nu_{(\alpha\beta)} = x_\alpha T_\beta^{*\mu} - x_\beta T_\alpha^{*\mu}$$

and

$$\partial_\mu J^{*\mu}_{(\alpha\beta)} = T_\beta^{*\mu} \eta_{\alpha\mu} - T_\alpha^{*\mu} \eta_{\beta\mu},$$

noting that $\partial_\mu x_\alpha = \eta_{\alpha\beta} \delta^\beta_\mu = \eta_{\alpha\mu}$. The last vanishes if and only if the matrix with entries

$$S_{\alpha\beta} = \eta_{\alpha\mu} T_\beta^{*\mu}$$

is *symmetric*, that is $S_{\alpha\beta} = S_{\beta\alpha}$. But this is *false*. Therefore the above argument does not yield a conservation law corresponding to spacetime rotations. This is not surprising: The naive point of view fails for rotations, because rotations, in contrast to translations, bring the tensor character of g into play.

We shall presently discuss Noether's theorem from a global geometric point of view. Consider maps

$$u\colon M \to \mathcal{N}$$

with $\dim M = n$ and $\dim \mathcal{N} = m$. As in Section 2.2, we denote by $\mathcal{V}$ the velocity bundle

$$\mathcal{V} = \bigcup_{(x,q)} \mathcal{L}(T_x M, T_q \mathcal{N}),$$

where $u(x) = q \in \mathcal{N}$ and $du(x) = v \in \mathcal{L}(T_x M, T_q \mathcal{N})$. A diffeomorphism f of M onto itself induces a diffeomorphism f_* of $\mathcal{V}$ onto itself by

$$v \in \mathcal{L}(T_x M, T_q \mathcal{N}) \mapsto f_*(v) \in \mathcal{L}(T_{f(x)} M, T_q \mathcal{N}), \tag{98}$$

where

$$f_*(v) \cdot Y = v \cdot (df^{-1} \cdot Y) \quad \forall\, Y \in T_{f(x)} M. \tag{99}$$

If X is a vector field on M generating the 1-parameter group $\{f_t\}$, we call the induced 1-parameter group $\{f_{t*}\}$ of diffeomorphisms of $\mathcal{V}$ the *Lie flow* generated by X on $\mathcal{V}$. Recall that a Lagrangian L is a section of the bundle $\pi^*_{\mathcal{V},M} \Lambda_n M$ over $\mathcal{V}$, the pullback by the projection $\pi_{\mathcal{V},M}\colon \mathcal{V} \to M$ of the bundle $\Lambda_n M$ of top-degree-forms on M. The pullback f^* by f of L is defined by

$$(f^* L)(v) \cdot (Y_1, \dots, Y_n) = L(f_*(v)) \cdot (df \cdot Y_1, \dots, df \cdot Y_n) \tag{100}$$

for all $v \in \mathcal{L}(T_x M, T_q \mathcal{N})$ and all $Y_1, \dots, Y_n \in T_x M$. (Recall that $f_*(v) \in \mathcal{L}(T_{f(x)} M, T_q \mathcal{N})$.) The Lie derivative $\mathcal{L}_X L$ of L with respect to a vector field X on M is the derivative of L with respect to the Lie flow generated by X on $\mathcal{V}$:

$$\mathcal{L}_X L = \frac{d}{dt} f_t^* L \Big|_{t=0}, \tag{101}$$

where $\{f_t\}$ is the 1-parameter group generated by X. A *current* J is a section of $\pi^*_{\mathcal{V},M} \Lambda_{n-1} M$. Given a volume form ϵ on M we consider J as equivalent to J^*, a section of $\pi^*_{\mathcal{V},M} TM$, as follows:

$$J(v) \cdot (Y_1, \dots, Y_{n-1}) = \epsilon(J^*(v), Y_1, \dots, Y_{n-1}), \tag{102}$$

for all $v \in \mathcal{L}(T_x M, T_q \mathcal{N})$ and all $Y_1, \dots, Y_{n-1} \in T_x M$. We now refer to the following notion from [12].

Definition 36. Given a Lagrangian L, we say that a current J is compatible with L, if there exists a section K of $\pi^*_{\mathcal{V},M} \Lambda_n M$ such that for every solution u of the Euler–Lagrange equations corresponding to L we have

$$d\,(J \circ du) = K \circ du. \tag{103}$$

In the above definition $J \circ du$ is an $(n-1)$-form on M while $K \circ du$ is an n form on M.

An interesting question is the following: What are all the possible currents compatible with a given Lagrangian? In $n = 1$ dimension all currents are compatible. However, the question becomes non-trivial in $n > 1$ dimensions. In fact, the Noether theorem provides a class of compatible currents in any dimension n of the domain manifold M.

Definition 37. The domain Noether current corresponding to a Lagrangian L and a vector field X on M is the current J, given by

$$J^{*\mu} = T_\nu^{*\mu} X^\nu, \tag{104}$$

where $T_\nu^{*\mu}$ are the components of the canonical stress. Recall that $T_\nu^{*\mu} = p_a^{*\mu} v_\nu^a - L^* \delta_\nu^\mu$ and $L(v) = L^*(v)\epsilon(x)$ while $p_a^{*\mu} = \frac{\partial L^*}{\partial v_\mu^a}$ are the components of the canonical momentum.

We can now state Noether's theorem in the domain case.

Theorem 5 (Noether's theorem in the domain case). *The Noether current is a compatible current and the corresponding section K of $\pi_{\mathcal{V},M}^* \Lambda_n M$ is given by*

$$K = -\mathcal{L}_X L. \tag{105}$$

In particular, if X generates a Lie flow leaving the Lagrangian invariant, then J is conserved, that is for every solution u of the Euler–Lagrange equations we have

$$d(J \circ du) = 0. \tag{106}$$

Hence, for two homologous hypersurfaces Σ_1 and Σ_2 it holds that

$$\int_{\Sigma_2} J \circ du = \int_{\Sigma_1} J \circ du. \tag{107}$$

Sections of Tensor Bundles over M. Having in mind the application to General Relativity we consider the case of the bundle $S_2 M$ of 2-covariant symmetric tensors on M. In fact, the configuration space $\mathcal{C}$ in General Relativity is the open subbundle LM of Lorentzian metrics on M:

$$\mathcal{C} = LM = \bigcup_{x \in M} L_x M, \tag{108}$$

where $L_x M$ is the set of quadratic forms on $T_x M$ of index 1, an open subset of the space $S_{2x} M$ of quadratic forms on $T_x M$. The velocity bundle $\mathcal{V}$ is

$$\mathcal{V} = \bigcup_{x \in M} \mathcal{L}(T_x M, S_{2x} M).$$

The Lagrangian is defined on the bundle product

$$\mathcal{C} \times_M \mathcal{V} = \bigcup_{x \in M} L_x M \times \mathcal{L}(T_x M, S_{2x} M).$$

(Note that $\dim(\mathcal{C} \times_M \mathcal{V}) = \dim \mathcal{C} + \dim \mathcal{V} - \dim M = n + \frac{n(n+1)}{2} + \frac{n^2(n+1)}{2}$.)

Given a diffeomorphism f of M onto itself we shall define the action of f on $\mathcal{C}$, $\mathcal{V}$ and on $\mathcal{C} \times_M \mathcal{V}$. Consider a section s of $\mathcal{C}$ (over M). Thus, s is a Lorentzian metric on M. We must associate to s a derived section which is a section of $\mathcal{V}$ (over M). Here, a connection Γ (on TM) is needed. We assume that Γ is symmetric. The derived section is then ∇s, the covariant derivative of s with respect to Γ. To represent things in terms of duals, we must also choose a volume form ϵ on M. This must be compatible with Γ, that is we must have

$$\nabla \epsilon = 0.$$

Such a choice is possible if and only if

$$\operatorname{tr} R(X, Y) = 0 \quad \forall\, X, Y \in T_x M, \ \forall x \in M,$$

where $R(X, Y)$ is the curvature transformation associated to Γ. (In an arbitrary local frame this condition reads $R^{\mu}_{\mu\alpha\beta} = 0$.) We now give the following definitions:

Definition 38. The action of f on $\mathcal{C}$ is defined by

$$q \in \mathcal{C}_x \mapsto f_*(q) \in \mathcal{C}_{f(x)}, \quad x \in M, \tag{109}$$

where

$$f_*(q) \cdot (Y_1, Y_2) = q \cdot (df^{-1} \cdot Y_1, df^{-1} \cdot Y_2) \quad \text{for all } Y_1, Y_2 \text{ in } T_{f(x)} M.$$

Definition 39. The action of f on $\mathcal{V}$ is defined by

$$v \in \mathcal{V}_x = \mathcal{L}(T_x M, S_{2x} M) \mapsto f_*(v) \in \mathcal{V}_{f(x)} = \mathcal{L}(T_{f(x)} M, S_{2 f(x)} M), \quad x \in M, \tag{110}$$

where

$$f_*(v) \cdot Y = f_*(v \cdot (df^{-1} \cdot Y)) \quad \text{for all } Y \text{ in } T_{f(x)} M.$$

The Lagrangian L is a section of

$$\pi^*_{\mathcal{C} \times_M \mathcal{V}, M} \Lambda_n M.$$

Thus, if $(q, v) \in L_x M \times \mathcal{L}(T_x M, S_{2x} M)$, then

$$L(q, v) \in \Lambda_{nx} M,$$

where $\Lambda_{nx}M$ is the space of totally antisymmetric n-linear forms in T_xM. The pullback by f of L, namely f^*L, is given by

$$(f^*L)(q,v)\cdot(Y_1,\dots,Y_n) = L(f_*(q), f_*(v))\cdot(df\cdot Y_1,\dots,df\cdot Y_n) \tag{111}$$

for all $Y_1,\dots,Y_n$ in T_xM. If X is a vector field on M generating the 1-parameter group $\{f_t\}$ of diffeomorphisms of M onto itself, the induced 1-parameter group $\{f_{t*}\}$ of diffeomorphisms of $\mathcal{C}\times_M \mathcal{V}$ onto itself, is the Lie flow generated by X on $\mathcal{C}\times_M\mathcal{V}$. It is actually generated by a vector field X_* on $\mathcal{C}\times_M\mathcal{V}$ which is expressed as the sum of its horizontal and vertical parts:

$$X_* = X_*^H + X_*^V. \tag{112}$$

We have

$$d\pi_{\mathcal{C}\times_M\mathcal{V},M}\cdot X_*^H = X \tag{113}$$

and

$$d\pi_{\mathcal{C}\times_M\mathcal{V},M}\cdot X_*^V = 0. \tag{114}$$

The horizontal part $X_*^H = X^\sharp$ is the horizontal lift of X to $\mathcal{C}\times_M\mathcal{V}$ defined by the connection Γ. The vertical part is given by

$$X^V(q,v) = \left(\frac{\nabla f_{t*}(q)}{\nabla t}\bigg|_{t=0}, \frac{\nabla f_{t*}(v)}{\nabla t}\bigg|_{t=0}\right),$$

(an element of $S_{2x}M\times\mathcal{L}(T_xM, S_{2x}M)$) which is a vector tangent to the fibre $\mathcal{C}_x\times\mathcal{L}(T_xM, S_{2x}M)$, an open set in the linear space $S_{2x}M\times\mathcal{L}(T_xM,S_{2x}M)$.[1] Here, $f_{t*}(q)$ is a field of Lorentzian tensors along the curve $f_t(x)$, namely the integral curve of X through x.

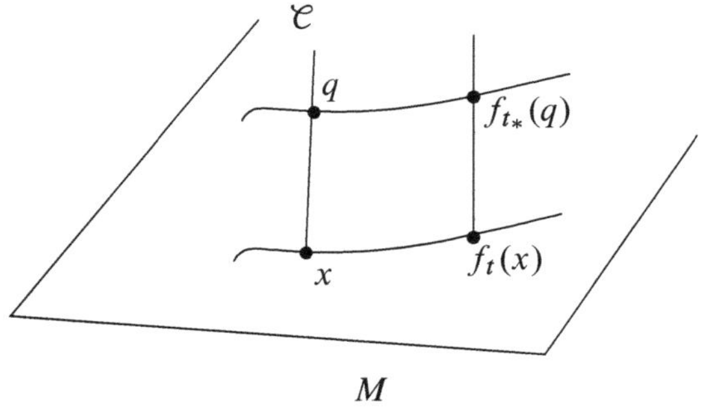

Figure 10

[1] Recall that a tangent vector at a point in a linear space can be thought of as an element of the linear space (a tangent vector at the origin).

Consider an arbitrary system of local coordinates on M. We can expand $q \in \mathcal{C}_x$ and $v \in \mathcal{L}(T_x M, S_{2x} M)$ in this system as

$$q = q_{\alpha\beta}\, dx^\alpha(x) \otimes dx^\beta(x),$$

where $q_{\beta\alpha} = q_{\alpha\beta}$, so q is symmetric, non-degenerate and of index 1. Moreover we can expand

$$v = v_{\mu\alpha\beta}\, dx^\mu(x) \otimes dx^\alpha(x) \otimes dx^\beta(x),$$

where $v_{\mu\beta\alpha} = v_{\mu\alpha\beta}$. Given $X \in T_x M$, we have $X = X^\alpha \frac{\partial}{\partial x^\alpha}\big|_x$ and

$$\begin{aligned} v \cdot X &= (v \cdot X)_{\alpha\beta}\, dx^\alpha(x) \otimes dx^\beta(x), \\ (v \cdot X)_{\alpha\beta} &= v_{\mu\alpha\beta} X^\mu. \end{aligned}$$

The $(q_{\alpha\beta})$ constitute a system of linear coordinates for $\mathcal{C}_x$ and the $(v_{\mu\alpha\beta})$ constitute a system of linear coordinates for $\mathcal{V}_x$. We can then express

$$X_*^V(q, v) = \left(\left(\frac{\nabla f_{t*}(q)}{\nabla t}\bigg|_{t=0} \right)_{\alpha\beta} \frac{\partial}{\partial q_{\alpha\beta}}, \left(\frac{\nabla f_{t*}(v)}{\nabla t}\bigg|_{t=0} \right)_{\mu\alpha\beta} \frac{\partial}{\partial v_{\mu\alpha\beta}} \right),$$

where

$$\begin{aligned} \left(\frac{\nabla f_{t*}(q)}{\nabla t}\bigg|_{t=0} \right)_{\alpha\beta} &= -q_{\gamma\beta} \nabla_\alpha X^\gamma(x) - q_{\alpha\gamma} \nabla_\beta X^\gamma(x), \\ \left(\frac{\nabla f_{t*}(v)}{\nabla t}\bigg|_{t=0} \right)_{\mu\alpha\beta} &= -v_{\nu\alpha\beta} \nabla_\mu X^\nu(x) - v_{\mu\nu\beta} \nabla_\alpha X^\nu(x) - v_{\mu\alpha\nu} \nabla_\beta X^\nu(x). \end{aligned}$$

In conclusion, for any differentiable function F on $\mathcal{C} \times_M \mathcal{V}$ we have

$$\begin{aligned} X_* F = X^\sharp F &- (q_{\gamma\beta} \nabla_\alpha X^\gamma + q_{\alpha\gamma} \nabla_\beta X^\gamma) \\ &\cdot \frac{\partial F}{\partial q_{\alpha\beta}} - (v_{\nu\alpha\beta} \nabla_\mu X^\nu + v_{\mu\nu\beta} \nabla_\alpha X^\nu + v_{\mu\alpha\nu} \nabla_\beta X^\nu) \frac{\partial F}{\partial v_{\mu\alpha\beta}}. \end{aligned}$$

The derivative of L with respect to the Lie flow generated by X is

$$\mathcal{L}_X L = \frac{d\, f_t^* L}{dt}\bigg|_{t=0}, \tag{115}$$

where $L = \frac{1}{n!} \sum L_{\lambda_1 \ldots \lambda_n}\, dx^{\lambda_1} \wedge \cdots \wedge dx^{\lambda_n}$. We have

$$(\mathcal{L}_X L)_{\lambda_1 \ldots \lambda_n} = X_*(L_{\lambda_1 \ldots \lambda_n}) + \sum_{i=1}^n \frac{\partial X^\kappa}{\partial x^{\lambda_i}} L_{\lambda_1 \cdots <\lambda_i> \kappa \ldots \lambda_n}.$$

Here, by $< \lambda_i > \kappa$ we mean that in the ith place the suffix λ_i is missing and in its place we have the suffix κ. Writing $L = L^* \epsilon$, where ϵ is a differentiable volume form on M ($\nabla \epsilon = 0$), we have

$$(\mathcal{L}_X L)^* = X_*(L^*) + (\nabla_\mu X^\mu) L^*. \tag{116}$$

We now give the following definition.

Definition 40. In a theory of Lorentzian metrics, the Noether current corresponding to a vector field X is

$$J^{*\mu}(q,v) = p^{*\mu\alpha\beta}(X^\nu v_{\nu\alpha\beta} + q_{\nu\beta}\nabla_\alpha X^\nu + q_{\alpha\nu}\nabla_\beta X^\nu) - L^* X^\mu. \tag{117}$$

The Euler–Lagrange equations are, in terms of the canonical momentum $p^{*\mu\alpha\beta} = \frac{\partial L^*}{\partial v_{\mu\alpha\beta}}$ and the canonical force $f^{*\alpha\beta} = \frac{\partial L^*}{\partial q_{\alpha\beta}}$,

$$\nabla_\mu \left(p^{*\mu\alpha\beta} \circ (s, \nabla s)\right) = f^{*\alpha\beta} \circ (s, \nabla s). \tag{118}$$

We then compute

$$\begin{aligned}
&\nabla_\mu \left(J^{*\mu} \circ (s, \nabla s)\right)\\
&\quad = \nabla_\mu \left(p^{*\mu\alpha\beta} \circ (s, \nabla s)\right) (X^\nu \nabla_\nu s_{\alpha\beta} + s_{\nu\beta}\nabla_\alpha X^\nu + s_{\alpha\nu}\nabla_\beta X^\nu)\\
&\qquad + p^{*\mu\alpha\beta} \circ (s, \nabla s) \{X^\nu \nabla_\mu \nabla_\nu s_{\alpha\beta} + s_{\nu\beta}\nabla_\mu\nabla_\alpha X^\nu + s_{\alpha\nu}\nabla_\mu\nabla_\beta X^\nu\\
&\qquad + (\nabla_\mu X^\nu)(\nabla_\nu s_{\alpha\beta}) + (\nabla_\mu s_{\nu\beta})(\nabla_\alpha X^\nu) + (\nabla_\mu s_{\alpha\nu})(\nabla_\beta X^\nu)\}\\
&\qquad - X^\sharp(L^*) - \frac{\partial L^*}{\partial q_{\alpha\beta}} \circ (s, \nabla s) \nabla_X s_{\alpha\beta}\\
&\qquad - \frac{\partial L^*}{\partial v_{\mu\alpha\beta}} \circ (s, \nabla s) \nabla_X(\nabla_\mu s_{\alpha\beta}) - (\nabla_\mu X^\mu) L^*,
\end{aligned}$$

and we have

$$\begin{aligned}
(\mathcal{L}_X L)^* = X^\sharp(L^*) &- \frac{\partial L^*}{\partial q_{\alpha\beta}} (q_{\nu\beta}\nabla_\alpha X^\nu + q_{\alpha\nu}\nabla_\beta X^\nu)\\
&- \frac{\partial L^*}{\partial v_{\mu\alpha\beta}} (v_{\nu\alpha\beta}\nabla_\mu X^\nu + v_{\mu\nu\beta}\nabla_\alpha X^\nu + v_{\mu\alpha\nu}\nabla_\beta X^\nu) + (\nabla_\mu X^\mu) L^*.
\end{aligned}$$

Hence, by the definitions of canonical momentum and force as well as the Euler–Lagrange equations,

$$\begin{aligned}
\nabla_\mu &\left(J^{*\mu} \circ (s, \nabla s)\right) + (\mathcal{L}_X L)^* \circ (s, \nabla s)\\
&= p^{*\mu\alpha\beta} \circ (s, \nabla s) \{X^\nu \nabla_\mu \nabla_\nu s_{\alpha\beta} - \nabla_X \nabla_\mu s_{\alpha\beta}\\
&\qquad + s_{\nu\beta}\nabla_\mu\nabla_\alpha X^\nu + s_{\alpha\nu}\nabla_\mu\nabla_\beta X^\nu\}.
\end{aligned} \tag{119}$$

We have

$$X^\nu \left(\nabla_\mu\nabla_\nu s_{\alpha\beta} - \nabla_\nu\nabla_\mu s_{\alpha\beta}\right) = -X^\nu \left(R^\kappa_{\alpha\mu\nu} s_{\kappa\beta} + R^\kappa_{\beta\mu\nu} s_{\alpha\kappa}\right).$$

Let us recall that

$$\nabla_\mu\nabla_\nu X^\lambda + R^\lambda_{\nu\kappa\mu} X^\kappa = (\mathcal{L}_X \Gamma)^\lambda_{\mu\nu}.$$

Consequently, the right-hand side of (119) is

$$p^{*\mu\alpha\beta} \circ (s, \nabla s) \{s_{\nu\beta} \,(\mathcal{L}_X \Gamma)^{\nu}_{\mu\alpha} + s_{\alpha\nu} \,(\mathcal{L}_X \Gamma)^{\nu}_{\mu\beta}\}. \tag{120}$$

In fact, the *Lie derivative* of a connection Γ is defined by:

$$(\mathcal{L}_X \Gamma)\,(Y, Z) = [X, \nabla_Y Z] - \nabla_{[X,Y]} Z - \nabla_Y [X, Z] \tag{121}$$

for any three vector fields X, Y, Z. One can readily check that $(\mathcal{L}_X \Gamma)\,(Y, Z)$ is bilinear in Y and Z with respect to multiplication by the ring of differentiable functions. (That is, we have $(\mathcal{L}_X \Gamma)(fY, Z) = f(\mathcal{L}_X \Gamma)\,(Y, Z)$ and $(\mathcal{L}_X \Gamma)(Y, fZ) = f(\mathcal{L}_X \Gamma)\,(Y, Z)$, for any differentiable function f.) Moreover, the fact that Γ is a symmetric connection implies that $(\mathcal{L}_X)\Gamma\,(Y, Z)$ is symmetric in Y, Z. It follows that $\mathcal{L}_X \Gamma$ is a T^1_2 tensor field which is symmetric in the lower indices. Since we have $[U, V] = \nabla_U V - \nabla_V U$ for any pair of vector fields U, V, (121) is equivalent to

$$\begin{aligned}(\mathcal{L}_X \Gamma)\,(Y, Z) &= \nabla_X \nabla_Y Z - \nabla_{\nabla_Y Z} X - \nabla_{[X,Y]} Z - \nabla_Y \nabla_X Z + \nabla_Y \nabla_Z X \\ &= R(X, Y)Z + \nabla^2 X\,(Y, Z).\end{aligned} \tag{122}$$

Substituting a frame field, we obtain

$$(\mathcal{L}_X \Gamma)^{\nu}_{\mu\lambda} = R^{\nu}_{\lambda\kappa\mu} X^{\kappa} + \nabla_\mu \nabla_\lambda X^{\nu}, \tag{123}$$

in agreement with the formula above.

We conclude from the above that the following form of Noether's theorem holds for sections of $\mathcal{C}$:

Theorem 6 (Noether's theorem for sections of $\mathcal{C}$). *The Noether current (from Definition* 40*) is a compatible current, that is*

$$d\ \ (J \circ (s, \nabla s)) = K \circ (s, \nabla s), \tag{124}$$

and we have

$$K = -\mathcal{L}_X L - T, \tag{125}$$

where

$$T^{*\mu} = -p^{\mu\alpha\beta}\ \{q_{\nu\beta}(\mathcal{L}_X \Gamma)^{\nu}_{\mu\alpha} + q_{\alpha\nu}(\mathcal{L}_X \Gamma)^{\nu}_{\mu\beta}\}. \tag{126}$$

In particular, if X generates a flow on M leaving Γ invariant as well as a Lie flow on $\mathcal{C} \times_M \mathcal{V}$ leaving L invariant, then for every solution s of the Euler–Lagrange equations we have

$$d\ \ (J \circ (s, \nabla s)) = 0. \tag{127}$$

We are now going to apply the above to General Relativity.

We introduce a background metric $\overset{\circ}{g}$ (in addition to the actual metric g). The background metric is not to be varied in the variational principle. To $\overset{\circ}{g}$ is associated its metric connection $\overset{\circ}{\Gamma}$ and volume form $\epsilon = d\mu_{\overset{\circ}{g}}$ (which is of course $\overset{\circ}{\Gamma}$-compatible). We write

$$-\frac{1}{4}\,(R - g^{\mu\nu}\ \overset{\circ}{R}_{\mu\nu})\, d\mu_g = (L^* - \overset{\circ}{\nabla}_\mu\ I^\mu)\, d\mu_{\overset{\circ}{g}}. \tag{128}$$

Defining the positive function ω by

$$d\mu_g = \omega\ d\mu_{\overset{\circ}{g}}, \tag{129}$$

L^* is given by

$$L^* = -\frac{1}{4}\omega\ g^{\mu\nu}\ (\Delta^\beta_{\mu\alpha}\Delta^\alpha_{\nu\beta} - \Delta^\beta_{\mu\nu}\Delta^\alpha_{\beta\alpha}), \tag{130}$$

where Δ is the difference of the two connections:

$$\begin{aligned}\Delta^\alpha_{\mu\nu} &= \Gamma^\alpha_{\mu\nu} - \overset{\circ}{\Gamma}{}^\alpha_{\mu\nu}\\ &= \frac{1}{2}g^{\alpha\beta}\ (\overset{\circ}{\nabla}_\mu\ g_{\beta\nu} + \overset{\circ}{\nabla}_\nu\ g_{\beta\mu} - \overset{\circ}{\nabla}_\beta\ g_{\mu\nu}).\end{aligned} \tag{131}$$

Now L^* is a true Lagrangian, as it depends only on g and $\overset{\circ}{\nabla}\ g$. Moreover,

$$I^\alpha = \frac{1}{4}\ \omega\ (g^{\mu\nu}\ \Delta^\alpha_{\mu\nu} - g^{\mu\alpha}\Delta^\nu_{\mu\nu}). \tag{132}$$

The Euler–Lagrange equations corresponding to L^* coincide with the Einstein equations, provided that $\overset{\circ}{g}$ is itself a solution of the Einstein equations, that is

$$\overset{\circ}{R}_{\mu\nu} = 0. \tag{133}$$

We take $\overset{\circ}{g}$ to be flat from this point on. The background spacetime is then the Minkowski spacetime. Let the vector field X generate a 1-parameter group of isometries of $\overset{\circ}{g}$. Then X leaves $\overset{\circ}{\Gamma}$, as well as $d\mu_{\overset{\circ}{g}} = \epsilon$, invariant. Moreover, the Lie flow generated by X leaves L^* invariant,

$$X_*(L^*) = 0. \tag{134}$$

We can thus apply Noether's theorem to conclude that we have a conserved current J_X associated to X.

Now, the background Minkowski spacetime has a 10-parameter isometry group, the *Poincaré group*. In Minkowski spacetime there is a special preferred class of coordinates, the rectangular coordinates, in which the metric components are

$$\overset{\circ}{g}_{\mu\nu} = \eta_{\mu\nu} = \operatorname{diag}(-1,1,1,1), \tag{135}$$

and

$$d\mu_{\overset{\circ}{g}} = dx^0 \wedge dx^1 \wedge dx^2 \wedge dx^3.$$

Moreover, it holds (as in all linear coordinates) that

$$\overset{\circ}{\Gamma}{}^{\alpha}_{\mu\nu} = 0. \tag{136}$$

The generators of the *Poincaré group* are expressed in rectangular coordinates by

$X_{(\alpha)} = \dfrac{\partial}{\partial x^\alpha}$ generators of spacetime translations along the x^α-coordinate axis,

$X_{(\alpha\beta)} = x_\alpha \dfrac{\partial}{\partial x^\beta} - x_\beta \dfrac{\partial}{\partial x^\alpha}$ generators of spacetime rotations in the (x^α, x^β)-coordinate plane, where $x_\alpha = \eta_{\alpha\beta} x^\beta$.

In terms of components in rectangular coordinates,

$$X^\mu_{(\alpha)} = \delta^\mu_\alpha,$$
$$X^\mu_{(\alpha\beta)} = x_\alpha \delta^\mu_\beta - x_\beta \delta^\mu_\alpha.$$

We therefore have the *conserved* Noether currents:

$J_{(\alpha)}$ associated to the translations $X_{(\alpha)}$,
$J_{(\alpha\beta)}$ associated to the rotations $X_{(\alpha\beta)}$.

Substituting in Definition 40 the generators of translations along the x^α-coordinate axis we obtain

$$J^{*\mu}_{(\alpha)} = T^{*\mu}_{(\alpha)}. \tag{137}$$

The associated conservation laws coincide with the Einstein–Weyl energy-momentum conservation laws, discussed in the preceding. But whereas the Einstein–Weyl approach provides no conservation law corresponding to spacetime rotations, the present approach does.

Consider the Noether current J_X associated to a vector field X generating a 1-parameter group of isometries of the background Minkowski metric $\overset{\circ}{g}$. We have shown above that J_X is conserved. This means that for every solution g of the Euler–Lagrange equations (the Einstein equations) we have

$$d\,(J_X \circ (g, \overset{\circ}{\nabla} g)) = 0, \tag{138}$$

which implies that if Σ_1 and Σ_2 are homologous hypersurfaces (in particular $\partial\Sigma_2 = \partial\Sigma_1$), then the following holds:

$$\int_{\Sigma_2} J_X \circ (g, \overset{\circ}{\nabla} g) = \int_{\Sigma_1} J_X \circ (g, \overset{\circ}{\nabla} g). \tag{139}$$

Next we are going to show that in fact J_X is a *boundary current*. In the general setting of a Lagrangian theory of maps u of the domain manifold M into another manifold $\mathcal{N}$, we have the following definition.

Definition 41. A current J is called a boundary current if there exists a section G of $\pi^*_{\mathcal{V},M}\Lambda_{n-2}M$ (with $\dim M = n$), such that for every solution u of the Euler–Lagrange equations we have

$$J \circ du = d\,(G \circ du). \tag{140}$$

A boundary current is a *conserved* current; in fact

$$\int_{\Sigma} J \circ du = \int_{\partial\Sigma} G \circ du. \tag{141}$$

The integral conservation law

$$\int_{\Sigma_2} J \circ du = \int_{\Sigma_1} J \circ du$$

trivially follows from the fact that $\partial\Sigma_2 = \partial\Sigma_1$ whenever Σ_2 is homologous to Σ_1.

Let us go back to the definition of the Noether current corresponding to a vector field X in a Lagrangian theory of Lorentzian metrics on M. In rectangular coordinates of the background Minkowski metric $\overset{\circ}{g}$,

$$\overset{\circ}{g}_{\mu\nu} = \eta_{\mu\nu}, \quad \overset{\circ}{\nabla}_\mu = \partial_\mu, \quad \det \overset{\circ}{g} = -1,$$

and we have

$$J_X^{*\mu} = T^{*\mu}_{\nu} X^\nu + p^{*\mu\alpha\beta}\,(g_{\nu\beta}\partial_\alpha X^\nu + g_{\alpha\nu}\partial_\beta X^\nu). \tag{142}$$

Taking X to be a Killing field of $\overset{\circ}{g}$, we have

$$X^\mu = \begin{cases} X^\mu_{(\alpha)} = \delta^\mu_\alpha & \text{for the generators of spacetime translations,} \\ X^\mu_{(\alpha\beta)} = x_\alpha \delta^\mu_\beta - x_\beta \delta^\mu_\alpha,\ x_\mu = \eta_{\mu\nu}x^\nu & \text{for the generators of spacetime rotations.} \end{cases}$$

It holds then that

$$\partial_\mu \partial_\nu X^\alpha = 0,$$

which is the equation

$$\mathcal{L}_X \overset{\circ}{\Gamma} = 0.$$

Now, in General Relativity it turns out that we have, identically,

$$T_\nu^{*\mu} = \partial_\lambda F_\nu^{*\mu\lambda} - \frac{1}{2}\sqrt{-\det g}\ \hat{R}_\nu^\mu, \tag{143}$$

where $F_\nu^{*\mu\lambda}$ is an expression which is *antisymmetric* in μ and λ, to be given below. (Recall that $\hat{R}_{\mu\nu} = R_{\mu\nu} - \frac{1}{2}g_{\mu\nu}R$ and $\hat{R}_\nu^\mu = g^{\mu\lambda}\hat{R}_{\lambda\nu}$.) Fixing the index ν let us write

$$T_{(\nu)\alpha\beta\gamma} = T_\nu^{*\mu}\ \overset{\circ}{\epsilon}_{\mu\alpha\beta\nu} \tag{144}$$

and

$$F_{(\nu)\alpha\beta} = \frac{1}{2} F_\nu^{*\mu\lambda}\ \overset{\circ}{\epsilon}_{\mu\lambda\alpha\beta}\,. \tag{145}$$

Then by (143) for every solution g of the Einstein equations we have

$$T_{(\nu)} = d\ F_{(\nu)}, \tag{146}$$

when composed with $(g, \partial g)$. The $F_\nu^{*\mu\lambda}$ are given by

$$F_\nu^{*\mu\lambda} = A_\nu^{\mu\lambda\kappa\alpha\beta}\partial_\kappa g_{\alpha\beta}, \tag{147}$$

where

$$\begin{aligned} A_\nu^{\mu\lambda\kappa\alpha\beta} = \frac{\sqrt{-\det g}}{8}\{ &\delta_\nu^\alpha\ (g^{\mu\kappa}g^{\lambda\beta} - g^{\lambda\kappa}g^{\mu\beta}) + \delta_\nu^\beta\ (g^{\mu\kappa}g^{\lambda\alpha} - g^{\lambda\kappa}g^{\mu\alpha}) \\ &+ 2g^{\alpha\beta}\ (\delta_\nu^\mu g^{\lambda\kappa} - \delta_\nu^\lambda g^{\mu\kappa}) + g^{\kappa\alpha}\ (g^{\mu\beta}\delta_\nu^\lambda - g^{\lambda\beta}\delta_\nu^\mu) \\ &+ g^{\kappa\beta}\ (g^{\mu\alpha}\delta_\nu^\lambda - g^{\lambda\alpha}\delta_\nu^\mu)\}. \end{aligned} \tag{148}$$

Using (143) we deduce that for any Killing field X of $\overset{\circ}{g}$ and every solution of the Einstein equations we have

$$J_X^{*\mu} = \partial_\lambda G_X^{*\mu\lambda}, \tag{149}$$

where

$$G_X^{*\mu\lambda} = F_\nu^{*\mu\lambda}X^\nu + K_X^{*\mu\lambda},$$

and $K_X^{*\mu\lambda}$ is an expression which is *antisymmetric* in μ and λ:

$$K_X^{*\mu\lambda} = \frac{1}{4}\{\ (\sqrt{-g}\ g^{\mu\kappa} - \eta^{\mu\kappa})\ \partial_\kappa X^\lambda - (\sqrt{-g}\ g^{\lambda\kappa} - \eta^{\lambda\kappa})\ \partial_\kappa X^\mu\ \}.$$

In fact we have the identity

$$J_X^{*\mu} = \partial_\lambda G_X^{*\mu\lambda} - \frac{1}{2}\sqrt{-\det g}\, \hat{R}^\mu_\nu X^\nu. \tag{150}$$

Thus, for any Killing field X of $\overset{\circ}{g}$ and every solution of the Einstein equations we have

$$J_X = dG_X,$$

where

$$(G_X)_{\alpha\beta} = \frac{1}{2} G^{*\mu\lambda}\, \overset{\circ}{\epsilon}_{\mu\lambda\alpha\beta}\ .$$

We shall now investigate the invariance properties of integrated quantities under *gauge transformations*, to be explained presently. Remark that we are free to pull-back the metric g by an arbitrary (orientation preserving) diffeomorphism f of M while keeping the background Minkowski metric $\overset{\circ}{g}$ fixed:

$$g \ \mapsto \ f^* g,$$

where

$$(f^*g)_{\mu\nu}(x) = \frac{\partial f^\alpha}{\partial x^\mu}(x)\frac{\partial f^\beta}{\partial x^\nu}(x) g_{\alpha\beta}(f(x)).$$

We consider the quantity Q associated to X, a given Killing field of $\overset{\circ}{g}$, and to S, a 2-surface which is homologous to 0 ($S = \partial\Sigma$ for some bounded hypersurface Σ):

$$Q_X(S) = \int_S G_X, \tag{151}$$

with $G_X = G_X(g, \partial g)$ defined above. Here, S and X are part of the background Minkowski structure. They are not affected by f. So, f acts like a gauge transformation. Consider then a 1-parameter group $\{f_t\}$ generated by a vector field Y.

$$\dot{g} = \frac{d}{dt}\, f_t^* g\Big|_{t=0} = \mathcal{L}_Y g.$$

We investigate

$$\dot{Q}_X(S) = \int_S \dot{G}_X.$$

If $\dot{Q}_X(S)$ were to vanish for all vector fields Y, then $Q_X(S)$ would be a *geometric invariant.* This would be so if $\dot{G}_X$ were of the form

$$\dot{G}_X = dI_X,$$

with I_X a 1-form. For, we would then have

$$\int_S \dot{G}_X = 0$$

as $\partial S = \emptyset$. However, this is unfortunately *false*.

Nevertheless, we shall presently show that, in the case that X is a spacetime translation, if g is asymptotic to $\overset{\circ}{g}$ at spacelike infinity in an appropriate manner, then $\dot{G}_X$ is in fact asymptotically of the form dI_X. The case that X is a spacetime rotation shall be treated later.

Setting $\xi_\mu = g_{\mu\nu} Y^\nu$ we have

$$\begin{aligned}\dot{g}_{\mu\nu} = (\mathcal{L}_Y g)_{\mu\nu} &= \nabla_\mu \xi_\nu + \nabla_\nu \xi_\mu \\ &= Y^\lambda \partial_\lambda g_{\mu\nu} + g_{\lambda\nu} \partial_\mu Y^\lambda + g_{\mu\lambda} \partial_\nu Y^\lambda.\end{aligned} \tag{152}$$

The last equation holds in any system of coordinates, in particular, in the rectangular coordinates of $\overset{\circ}{g}$.

Now let $g_{\mu\nu} - \eta_{\mu\nu} = o_2(r^{-\alpha})$. Here, $r = \sqrt{\sum_{i=1}^3 (x^i)^2}$ and we say that a function f of the spacetime coordinates is $o_k(r^{-\alpha})$ if f is C^k and its partial derivatives with respect to the spacetime coordinates of order l are $o(r^{-\alpha-l})$, for all $l = 0, \dots, k$. We are only allowing 'gauge transformations' which do not affect this fall-off property. That is, we assume that

$$\xi_\mu = o_3(r^{1-\alpha}).$$

Then we have

$$\dot{g}_{\mu\nu} = \partial_\mu \xi_\nu + \partial_\nu \xi_\mu + o_1(r^{-2\alpha}). \tag{153}$$

Note that for translations $X_{(\nu)} = \frac{\partial}{\partial x^\nu}$,

$$G^{*\mu\lambda}_{X_{(\nu)}} = F^{*\mu\lambda}_\nu. \tag{154}$$

Now, we have

$$\dot{F}^{*\mu\lambda}_\nu = \underbrace{\dot{A}^{\mu\lambda\kappa\alpha\beta}_\nu \partial_\kappa g_{\alpha\beta}}_{o_1(r^{-1-2\alpha})} + A^{\mu\lambda\kappa\alpha\beta}_\nu \partial_\kappa \dot{g}_{\alpha\beta}, \tag{155}$$

hence

$$\dot{F}^{*\mu\lambda}_\nu = \overset{\circ}{A}{}^{\mu\lambda\kappa\alpha\beta}_\nu \partial_\kappa (\partial_\alpha \xi_\beta + \partial_\beta \xi_\alpha) + o_1(r^{-1-2\alpha}). \tag{156}$$

Let B_R be the largest coordinate ball ($r \le R$) contained in S. Then, if $\alpha \ge \frac{1}{2}$,

$$\int_S o(r^{-1-2\alpha}) = o(R^{1-2\alpha}) \to 0$$

as $R \to \infty$. It thus suffices to show that there is a 1-form $I_{(\nu)}$ such that

$$\frac{1}{2} \overset{\circ}{A}{}_{\nu}^{\mu\lambda\kappa\alpha\beta} \partial_\kappa \left(\partial_\alpha \xi_\beta + \partial_\beta \xi_\alpha\right) \overset{\circ}{\epsilon}_{\mu\lambda\gamma\delta} = \partial_\gamma I_{(\nu)\delta} - \partial_\delta I_{(\nu)\gamma}. \tag{157}$$

In fact, defining the totally antisymmetric expressions

$$I_{(\nu)}^{*\alpha\beta\gamma} = \frac{1}{4} \left\{ \delta_\nu^\alpha \left(\partial^\beta \xi^\gamma - \partial^\gamma \xi^\beta\right) + \delta_\nu^\beta \left(\partial^\gamma \xi^\alpha - \partial^\alpha \xi^\gamma\right) + \delta_\nu^\gamma \left(\partial^\alpha \xi^\beta - \partial^\beta \xi^\alpha\right) \right\}$$

we have

$$\overset{\circ}{A}{}_{\nu}^{\mu\lambda\kappa\alpha\beta} \partial_\kappa \left(\partial_\alpha \xi_\beta + \partial_\beta \xi_\alpha\right) = \partial_\rho I^{*\rho\mu\lambda}.$$

Hence, setting

$$I_{(\nu)\lambda} = \frac{1}{6} I_{(\nu)}^{*\alpha\beta\gamma} \overset{\circ}{\epsilon}_{\alpha\beta\gamma\lambda},$$

(157) indeed holds. We conclude the following:

Proposition 4. *Let S_r be the coordinate sphere of radius r on a complete asymptotically flat Cauchy hypersurface Σ. Then the limiting quantity*

$$Q_X(S_\infty) := \lim_{r\to\infty} Q_X(S_r) \tag{158}$$

associated to a translation X is a geometric invariant provided that $g_{\mu\nu} - \eta_{\mu\nu} = o_2(r^{-\alpha})$ for some $\alpha \geq \frac{1}{2}$.

3.2 Asymptotic flatness

What we have just shown implies that the *total energy-momentum* is a *geometric invariant*, provided that the metric $g_{\mu\nu}$ is asymptotically flat in the following sense: There is a coordinate system in the neighbourhood of spatial infinity with respect to which the metric components satisfy

$$g_{\mu\nu} = \eta_{\mu\nu} + o_2(r^{-\alpha}), \quad \alpha > \frac{1}{2}. \tag{159}$$

We shall now show that the energy E and the linear momentum P^j are well defined under the same assumption. The argument here follows [4]. Since we have shown these quantities to be independent of the particular choice of such a coordinate system, we may use coordinates such that the spatial coordinate lines ($x^i = c^i$, $i = 1, 2, 3$) are orthogonal to the hypersurfaces $x^0 = c^0$, that is we can set

$$g_{0i} = 0.$$

(We have $g_{00} = -\Phi^2$, $g_{ij} = \bar{g}_{ij}$.) We then find

$$F_0^{*0i} = -e^i, \tag{160}$$

where

$$e^i = \frac{\sqrt{\det \bar{g}}}{4} \left(\bar{g}^{jm}\bar{g}^{in} - \bar{g}^{ij}\bar{g}^{mn}\right) \partial_j \bar{g}_{mn}. \tag{161}$$

Thus, the energy is given by

$$E = \lim_{r\to\infty} \int_{S_r} e^i \, dS_i \tag{162}$$

(in agreement with the ADM expression). Also, we find

$$F_j^{*0i} = p_j^i = -\frac{\sqrt{\det \bar{g}}}{2} \left(\bar{g}^{im} k_{jm} - \delta_j^i \operatorname{tr} k\right), \tag{163}$$

where

$$p_j^i = -\frac{\sqrt{\det \bar{g}}}{2} \left(\bar{g}^{im} k_{jm} - \delta_j^i \operatorname{tr} k\right) \tag{164}$$

and $k_{ij} = \frac{1}{2\Phi}\frac{\partial \bar{g}_{ij}}{\partial t}$ is the 2nd fundamental form of the hypersurfaces $x^0 = c^0$. Thus, the linear momentum is given by

$$P^j = \lim_{r\to\infty} \int_{S_r} p_j^i \, dS_i \tag{165}$$

(in agreement with the ADM expression).

The hypotheses on Φ, $\bar{g}_{ij}$ and k_{ij} corresponding to (159) are

$$\begin{aligned} \Phi &= 1 + o_2(r^{-\alpha}), \\ \bar{g}_{ij} &= \delta_{ij} + o_2(r^{-\alpha}), \\ k_{ij} &= o_1(r^{-1-\alpha}), \quad \alpha > \frac{1}{2}. \end{aligned} \tag{166}$$

In the following, given a function f defined on a hypersurface $x^0 = t$, we shall take $f = o_k(r^{-\alpha})$ to mean that f is a C^k function of the spatial coordinates and its partial derivatives with respect to the spatial coordinates of order l are $o(r^{-\alpha-l})$, for all $l = 0, \dots, k$. Moreover, if a compact interval $[t_1, t_2]$ of values of t is considered, uniformity of the limit as $r \to \infty$ with respect to t is implied. In particular, the hypotheses (166) are to be meant in this sense.

We first show that under the hypotheses (166) the limit $r \to \infty$ in (162) and (165) exists. In fact we show a stronger result, namely that there exists a limit independent of the exhaustion. That is, we do not assume that the exhaustion of Σ is by concentric coordinate balls. Consider then nested domains B_n (i.e., $B_{n+1} \supset B_n$) such that $\bigcup_n B_n = \Sigma$ and suppose that the $S_n = \partial B_n$ are C^1. We claim that

$$\lim_{n\to\infty} \int_{S_n} e^i \, dS_i = E, \tag{167}$$

$$\lim_{n\to\infty} \int_{S_n} p^i_j \, dS_i = P^j. \tag{168}$$

To show this recall that for the 4-dimensional spacetime manifold we have

$$\partial_\alpha I^\alpha = \frac{1}{4} R \sqrt{-\det g} + L^*.$$

The analogous formula for the 3-dimensional manifold $(\Sigma, \bar{g})$ is

$$\partial_i \bar{I}^i = \frac{1}{4} \bar{R} \sqrt{\det \bar{g}} + \bar{L}^*,$$

where in fact

$$\bar{I}^i = e^i.$$

Also, we have

$$\bar{L}^* = -\frac{1}{4} \bar{g}^{mn} (\bar{\Gamma}^j_{mi} \bar{\Gamma}^i_{nj} - \bar{\Gamma}^j_{mn} \bar{\Gamma}^i_{ji}).$$

We now appeal to the constraint equation

$$\bar{R} + (\operatorname{tr} k)^2 - |k|^2 = 0$$

(the twice contracted Gauss equation) to conclude that, under the hypotheses (166),

$$\bar{R} = o_1(r^{-2-2\alpha}).$$

Also, we have

$$\bar{L}^* = o_1(r^{-2-2\alpha}).$$

Consider then two domains B and B' with $B' \supset B$ such that B contains the coordinate ball of radius R. Then we have

$$\begin{aligned}\int_{S'} e^i \, dS_i - \int_S e^i \, dS_i &= \int_{B' \setminus B} \partial_i e^i \, d^3x \\ &\le C \int_{B^c} r^{-2-2\alpha} \, d^3x \\ &\le C R^{1-2\alpha} \to 0 \quad \text{as } R \to \infty,\end{aligned}$$

since $\alpha > \frac{1}{2}$. This establishes our claim in the case of E.

The case of P^j is similar. Writing

$$\partial_i \, p^i_j = -\frac{1}{2}\, \partial_i \, \{\sqrt{\det \bar{g}}\, (\bar{g}^{im} k_{jm} - \delta^i_j \operatorname{tr} k)\},$$

we see that

$$\partial_i \, p^i_j = -\frac{1}{2}\, \sqrt{\det \bar{g}}\, \{\nabla_i \, (k^i_j - \delta^i_j \operatorname{tr} k) + \bar{\Gamma}^m_{ij} \, (k^i_m - \delta^i_m \operatorname{tr} k)\}.$$

We now appeal to the constraint equation

$$\nabla_i \, (k^i_j - \delta^i_j \operatorname{tr} k) = 0,$$

(the contracted Codazzi equation) to conclude that, under the hypotheses (166),

$$\begin{aligned} \partial_i \, p^i_j &= -\frac{1}{2}\, \sqrt{\det \bar{g}}\, \bar{\Gamma}^m_{ij} \, (k^i_m - \delta^i_m \operatorname{tr} k) \\ &= o(r^{-2-2\alpha}). \end{aligned}$$

Thus the same argument applies.

We proceed to discuss conservation of the total energy-momentum. Consider first the total energy. We have

$$E(t_2) - E(t_1) = \lim_{r\to\infty} \int_{t_1}^{t_2} \int_{S_r} \frac{\partial e^i}{\partial t} \, dS_i \, dt \tag{169}$$

and from (161),

$$\frac{\partial e^i}{\partial t} = \frac{\partial G^{ijmn}}{\partial t} \, \partial_j \bar{g}_{mn} + G^{ijmn} \, \partial_j \Big(\frac{\partial \bar{g}_{mn}}{\partial t} \Big),$$

where

$$G^{ijmn} = \frac{\sqrt{\det \bar{g}}}{8} \, (\bar{g}^{im} \bar{g}^{jn} + \bar{g}^{in} \bar{g}^{jm} - 2 \bar{g}^{ij} \bar{g}^{mn}).$$

Since

$$\frac{\partial \bar{g}_{ij}}{\partial t} = 2\, \Phi \, k_{ij},$$

we have

$$2\, G^{ijmn} \, \partial_j \, (\Phi k_{mn}) = 2\, G^{ijmn} \, (\underbrace{\Phi \partial_j k_{mn}}_{o(r^{-2-\alpha})} + \underbrace{(\partial_j \Phi) k_{mn}}_{o(r^{-2-2\alpha})}).$$

Thus we obtain

$$\frac{\partial e^i}{\partial t} = o(r^{-2-\alpha})$$

hence

$$r^2 \frac{\partial e^i}{\partial t} \to 0$$

uniformly as $r \to \infty$. Therefore we conclude that

$$E(t_2) - E(t_1) = 0. \tag{170}$$

Consider next the total linear momentum. We have

$$P^j(t_2) - P^j(t_1) = \lim_{r \to \infty} \int_{t_1}^{t_2} \int_{S_r} \frac{\partial p^i_j}{\partial t} \, dS_i \, dt \tag{171}$$

and from (164)

$$\frac{\partial p^i_j}{\partial t} = -\frac{1}{2} \sqrt{\det \bar{g}} \left\{ \Phi \operatorname{tr} k \, (k^i_j - \delta^i_j \operatorname{tr} k) + \frac{\partial k^i_j}{\partial t} - \delta^i_j \frac{\partial \operatorname{tr} k}{\partial t} \right\}. \tag{172}$$

(Note that $\frac{\partial \sqrt{\det \bar{g}}}{\partial t} = \Phi \operatorname{tr} k \sqrt{\det \bar{g}}$.) Appealing to the 2nd variational formula

$$\frac{\partial k_{ij}}{\partial t} = \bar{\nabla}_i \bar{\nabla}_j \Phi - \Phi \, (\bar{R}_{ij} - 2k_{im}k^m_j + k_{ij} \operatorname{tr} k),$$

and using the assumptions (166), we deduce that

$$\frac{\partial k_{ij}}{\partial t} = o(r^{-2-\alpha}).$$

Using this we obtain

$$\frac{\partial p^i_j}{\partial t} = o(r^{-2-\alpha});$$

hence

$$r^2 \frac{\partial p^i_j}{\partial t} \to 0$$

uniformly as $r \to \infty$. Therefore we conclude that

$$P^j(t_2) - P^j(t_1) = 0. \tag{173}$$

3.2.1 The maximal time function. Up to this point we have not made any particular choice of time function and this arbitrariness is reflected in the fact that the lapse function Φ is not subject to any equation.

We now require the level sets Σ_t of the time function t to be *maximal spacelike hypersurfaces*. That is, any compact perturbation of Σ_t decreases its volume. Thus Σ_t satisfies the maximal hypersurface equation

$$\operatorname{tr} k = 0. \tag{174}$$

In fact, for a spacetime which is asymptotically flat (in the sense above) and satisfying a certain barrier condition, there exists a *unique complete maximal hypersurface* asymptotic to a spacelike coordinate hyperplane at spatial infinity. This has been established by R. Bartnic in [3].

Definition 42. A maximal time function is a time function t whose level sets are maximal spacelike hypersurfaces which are complete and tend to parallel spacelike coordinate hyperplanes at spatial infinity. Moreover, the associated lapse function Φ is required to tend to 1 at spatial infinity.

Remark. We have one such function (up to an additive constant) for each choice of family of parallel spacelike hyperplanes in the background Minkowski spacetime. Two such families are related by the action of an element of the Lorentz group.

We now fix the family by requiring

$$P^i = 0, \tag{175}$$

that is, we require the total linear momentum to vanish. Then for any spacetime other than Minkowski spacetime we obtain a unique time function t (up to an additive constant) the *canonical maximal time function*. This is a consequence of the fact that any non-trivial spacetime has positive energy, which is the positive energy theorem, to be discussed in the next section. The choice (175) corresponds to the center-of-mass-frame in Newtonian mechanics.

Let us consider now the *Einstein equations relative to a maximal time function*,

$$\operatorname{tr} k = 0.$$

The constraint equations read

$$\text{Codazzi:} \qquad \nabla^j k_{ij} = 0, \tag{176}$$

$$\text{Gauss:} \qquad \bar{R} = |k|^2. \tag{177}$$

The evolution equations read

$$1^{\text{st}} \text{ variation:} \qquad \frac{\partial \bar{g}_{ij}}{\partial t} = 2\,\Phi\, k_{ij}, \tag{178}$$

$$2^{\text{nd}} \text{ variation:} \qquad \frac{\partial k_{ij}}{\partial t} = \bar{\nabla}_i \bar{\nabla}_j \Phi - (\bar{R}_{ij} - 2\, k_{im} k_j^m)\, \Phi. \tag{179}$$

Moreover, the trace of the 2nd variation equations yields by virtue of the maximality condition the *lapse equation*,

$$\bar{\triangle}\, \Phi - |k|^2\, \Phi = 0. \tag{180}$$

Choosing now the canonical maximal time function, we have

$$P_j = -\frac{1}{2} \lim_{r\to\infty} \int_{S_r} p^i_j \, dS_i = 0$$

with $p^i_j = k^i_j$. This allows us to impose stronger fall-off on k_{ij}. We thus introduce the notion of a *strongly asymptotically flat initial data set.*

Definition 43. A strongly asymptotically flat initial data set is an initial data set $(\overline{M}, \bar{g}, k)$ such that:

1. $\overline{M}$ is Euclidean at infinity.
2. There exists a coordinate system in the neighbourhood of infinity in $\overline{M}$ (that is on $\overline{M} \setminus \mathcal{K}$) in which the metric components satisfy
$$\bar{g}_{ij} = \left(1 + \frac{2M}{r}\right) \delta_{ij} + o_2(r^{-1}). \tag{181}$$
3. In the same coordinate system we have
$$k_{ij} = o_1(r^{-2}). \tag{182}$$

The total energy is then given by
$$E = 4\,\pi\, M. \tag{183}$$

3.2.2 Positivity of the energy. We shall now discuss the *positive energy theorem* of Schoen and Yau.

Theorem 7 (Positive energy theorem, Schoen–Yau [26]).

1. *Under the assumption*
$$\bar{g}_{ij} = \left(1 + \frac{2M}{r}\right) \delta_{ij} + O_2(r^{-2}), \tag{184}$$
it holds that
$$M \geq 0. \tag{185}$$
2. *If also the remainder is* $O_4(r^{-2})$, *then*
$$M = 0 \tag{186}$$
implies that $\bar{g}$ *is the Euclidean metric.*

Here and in the following $f = O_k(r^{-a})$ signifies that f is a C^* function and its partial derivatives of order l are $O(r^{-a-l})$ for all $l = 0, \dots, k$.

Proof. The theorem concerns strongly asymptotically Euclidean 3-manifolds $(\overline{M}, \bar{g})$ with non-negative scalar curvature $\overline{R} \geq 0$. To simplify the notation, we drop the overlines in the present section.

Proof of Part 1. $M \geq 0$. This consists of three steps.

Step 0: Given such g with $M < 0$, we can find another metric $\tilde{g}$ close enough to g so that also $\widetilde{M} < 0$ and such that $\widetilde{R} > 0$ everywhere.

Step 1: If $M < 0$, then barriers would exist for minimal surfaces which can be used to construct a complete area-minimizing minimal surface.

Step 2: Using the 2nd variation formula for area we show that the existence of the minimal surface in Step 1 contradicts the fact that $R > 0$ everywhere.

Step 0. Under a conformal change of g,

$$\tilde{g} = \Phi^4 \, g, \tag{187}$$

the scalar curvature $\widetilde{R}$ of $\tilde{g}$ is related to the scalar curvature R of g by

$$\widetilde{R} = \Phi^{-5} \, (R\Phi - 8\triangle\Phi). \tag{188}$$

Here $\triangle$ is the Laplacian of the metric g. Let us choose a smooth positive function f on $\overline{M}$ such that

$$f = O_2(r^{-4}).$$

We then wish to find Φ such that

$$\widetilde{R} = \Phi^{-4} \, R + \epsilon \, f,$$

where ϵ is a positive constant. We shall show that this is possible for suitably small ϵ. Now, by (188), Φ is subject to the equation

$$\triangle \, \Phi = -\frac{1}{8} \, \epsilon \, f \, \Phi^5.$$

Setting $\Phi = 1 + \Psi$, this equation reads

$$\triangle \, \Psi = -\frac{1}{8} \, \epsilon \, f \, (1 + \Psi)^5.$$

The implicit function theorem applies if ϵ is suitably small to give us the existence of a solution Ψ such that

$$\Psi = O_2(\epsilon \, r^{-1}).$$

In particular, $\Phi = 1 + \Psi$ is everywhere positive and

$$\tilde{M} = M + O(\epsilon) < 0, \tag{189}$$

if ϵ is suitably small. In fact, we have $\tilde{M} = M + 2N$, where

$$N = \lim_{r\to\infty}(r\Psi) = -\frac{1}{4\pi}\int_{\overline{M}} \Delta\Psi.$$

Step 1. This is based on the following lemma.

Lemma. *Let S be a C^2 minimal hypersurface (that is, a critical point for the area functional), in an n-dimensional smooth Riemannian manifold (M, g) $(\dim S = n-1)$, which is compact with boundary. Let f be a C^2 function on M such that the values λ, for $\lambda \in [\lambda_0, \infty)$, are non-critical and the corresponding level sets Σ_λ have positive mean curvature with respect to the unit normal pointing in the direction of increase of f. Then $f \le \lambda_0$ on ∂S implies $f \le \lambda_0$ on S.*

Proof of the lemma. We consider the restriction $\bar{f}$ of f to the hypersurface S. If the lemma is not true, then the subset $\mathcal{U}$ of S, where $\bar{f} > \lambda_0$ is non-empty, $\mathcal{U}$ is open; $\bar{f}$ then attains a maximum $\lambda_1 > \lambda_0$ at a point $p_M \in \mathcal{U}$. We have

$$(\overline{\nabla}\bar{f})(p_M) = 0$$

and

$$(\overline{\nabla}^2\bar{f})(p_M) \le 0.$$

Choosing a local frame field $(E_a,\ a = 1, \dots, n-1)$ for S in a neighbourhood of p_M in S, we write

$$\overline{\nabla}_a\overline{\nabla}_b\bar{f} \quad \text{for} \quad \overline{\nabla}^2 f \cdot (E_a, E_b).$$

We have

$$\begin{aligned}\overline{\nabla}_a\overline{\nabla}_b\bar{f} &= E_a(E_b\,\bar{f}) - (\overline{\nabla}_{E_a}E_b)\,\bar{f}\\ &= E_a(E_b\,\bar{f}) - (\Pi\nabla_{E_a}E_b)\,\bar{f}\end{aligned}$$

where Π is the orthogonal projection to S,

$$\Pi\nabla_{E_a}E_b = \nabla_{E_a}E_b - g(N, \nabla_{E_a}E_b)\,N,$$

with N the unit normal to S. Moreover, we have

$$\begin{aligned}g(N, \nabla_{E_a}E_b) &= -g(\nabla_{E_a}N, E_b)\\ &= -\theta_{ab}\end{aligned}$$

with θ the 2nd fundamental form of S. We thus arrive at the formula

$$\overline{\nabla}_a\overline{\nabla}_b\bar{f} = \nabla_a\nabla_b f - \theta_{ab}Nf \tag{190}$$

on S.

Next we take the trace of (190) with respect to the induced metric $\bar{g}$, $\bar{g}_{ab} = g(E_a, E_b)$, to obtain

$$\bar{\Delta}\bar{f} = \bar{g}^{ab}\nabla_a\nabla_b f, \tag{191}$$

recalling that $\operatorname{tr}\theta = 0$ as S is a minimal hypersurface. At an interior maximum point p_M, the Hessian matrix $(\bar{\nabla}_a\bar{\nabla}_b\bar{f})(p_M)$ is negative semi-definite. We thus have

$$(\bar{\Delta}\,\bar{f})(p_M) = (\bar{g}^{ab}\bar{\nabla}_a\bar{\nabla}_b\bar{f})(p_M) \leq 0. \tag{192}$$

Consider, on the other hand, the level sets Σ_λ of f in M. The unit normal vector field to Σ_λ, N', pointing in the direction of increase of f is

$$N'^i = g^{ij}\,\frac{\partial_j f}{|\nabla f|}.$$

Now, since p_M is a critical point of $\bar{f}$, the tangent planes $T_{p_M}S$ and $T_{p_M}\Sigma_{\lambda_1}$ *coincide*:

$$N'(p_M) = N(p_M).$$

Therefore, the induced metric is the same at p_M. Let $(E'_a, a = 1,\dots,n-1)$ be a local frame field for Σ_λ in a neighbourhood of p_M coinciding at p_M with the frame field $(E_a, a = 1,\dots,n-1)$ for S. Then the 2nd fundamental form of Σ_λ, θ'_{ab}, is given in this frame field by

$$\begin{aligned}
\theta'_{ab} &= g(\nabla_{E'_a}N', E'_b)\\
&= g_{ij}\;E'^k_a\;\nabla_k\Big(g^{il}\,\frac{\partial_l f}{|\nabla f|}\Big)\;E'^j_b\\
&= E'^k_a\;E'^l_b\;\frac{\nabla_k\nabla_l f}{|\nabla f|} - E'^k_a\,E'^l_b\;\partial_l f\;\frac{\partial_k|\nabla f|}{|\nabla f|^2}.
\end{aligned}$$

The last term vanishes. We thus obtain

$$\theta'_{ab} = \frac{\nabla^2 f\cdot(E'_a, E'_b)}{|\nabla f|}.$$

In particular, at p_M we have $E'_a = E_a$ hence

$$\theta'_{ab}(p_M) = \frac{(\nabla_a\nabla_b f)(p_M)}{|\nabla f(p_M)|}.$$

Therefore the following holds:

$$(\bar{g}^{ab}\;\nabla_a\nabla_b\;f)(p_M) = \operatorname{tr}\theta'(p_M)\;|\nabla f(p_M)| > 0. \tag{193}$$

In view of equation (191), (193) *contradicts* (192). This establishes the lemma. □

We now consider the circles C_σ with σ a constant on the coordinate plane $x^3 = 0$ in the neighbourhood of infinity on the manifold $(\overline{M}, g)$.

$$C_\sigma = \{(x^1, x^2, 0) : (x^1)^2 + (x^2)^2 = \sigma^2\}.$$

We can find a surface of least area S_σ spanning C_σ. Taking a sequence $\sigma_n \uparrow \infty$, and we want to find barriers which will allow us to conclude that we can extract a convergent subsequence and thus, passing to the limit, obtain a complete minimal surface S.

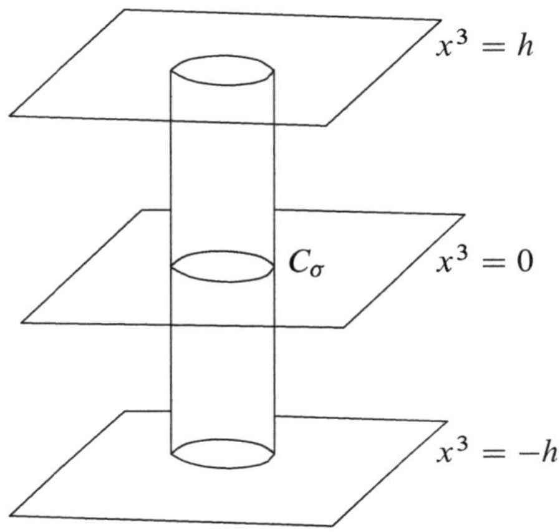

Figure 11

We apply the lemma to the functions $f = (x^1)^2 + (x^2)^2$ and $f = x^3$. In the first case the level sets are the coordinate cylinders $K_\sigma = \{(x^1, x^2, x^3) : (x^1)^2 + (x^2)^2 = \sigma^2\}$ and in the second case the level sets are the coordinate planes $x^3 = \lambda$, the P_λ. For K_σ the mean curvature is

$$\frac{1}{\sigma} + O(r^{-2}) > 0$$

for sufficiently large σ. For P_h and P_{-h}, where h is a positive constant, the mean curvature is

$$\frac{-2Mh}{r^3} + O(r^{-3}). \tag{194}$$

If $M < 0$, this is also > 0, provided that h is taken suitably large. In fact, the x_1 and x_2 are coordinates on the planes P_λ and we have

$$\begin{aligned} \nabla_a \nabla_b \, x^3 &= -\Gamma^3_{ab} \\ &= \frac{M}{r^3} (x^a \delta_{b3} + x^b \delta_{a3} - x^3 \delta_{ab}) + O(r^{-3}). \end{aligned}$$

Applying then the formula

$$\operatorname{tr}\theta' = \bar{g}^{ab}\,\frac{\nabla_a\nabla_b f}{|\nabla f|}$$

yields (194). Let

$$E_{\sigma,h} = \{(x^1,x^2,x^3) :\ (x^1)^2 + (x^2)^2 \le \sigma \text{ and} - h \le x_3 \le h\}.$$

By virtue of the lemma, the surfaces S_{σ_n} are contained in $E_{\sigma_n,h}$. We can then appeal to interior regularity theory for minimal surfaces to obtain uniform interior C^3 estimates and thus conclude that we can extract a subsequence $S_{\sigma_{n_i}}$ converging uniformly on compact domains to a complete C^2 area minimizing surface S. This completes the proof of Step 1.

Proof of Step 2. The surface constructed in Step 1 leads to a contradiction when considering the 2nd variational formula for area. A variation of a complete surface S is a 1-parameter family S_t such that $S_0 = S$. The family $\{S_t\}$ may not be a foliation. Nevertheless, the same approach as for a foliation applies if we consider a smooth mapping

$$h\colon (-\epsilon,\epsilon)\times S \to M.$$

The curves

$$h_p\colon (-\epsilon,\epsilon) \to M$$

with $h_p(t) = h(t,p)$ may not be orthogonal to the surfaces

$$S_t = h_t(S)$$

with $h_t(p) = h(t,p)$. We can nevertheless construct an orthogonal family and thus redefine the homotopy h. The following formulas hold relative to such a normalized homotopy:

$$\begin{aligned} A &= \int_S d\mu_{\bar{g}},\\ \frac{dA}{dt} &= \int_S f \operatorname{tr}\theta\, d\mu_{\bar{g}}. \end{aligned} \tag{195}$$

This is the first variation of the area. Here f is the *lapse function* (which measures the normal separation of the S_t), defined by

$$\frac{\partial}{\partial t} = fN,$$

where N is the unit normal to S_t. In fact the 1st variation equations

$$\frac{\partial \bar{g}_{ab}}{\partial t} = 2f\theta_{ab}$$

imply

$$\frac{\partial\, d\mu_{\bar g}}{\partial t} = f \operatorname{tr}\theta \, d\mu_{\bar g}.$$

The second variation of the area is then

$$\frac{d^2 A}{dt^2} = \int_S \left\{ f \frac{\partial \operatorname{tr}\theta}{\partial t} + (f \operatorname{tr}\theta)^2 + \frac{\partial f}{\partial t} \operatorname{tr}\theta \right\} d\mu_{\bar g}. \tag{196}$$

In our case S is a minimal surface.

Definition 44. A minimal surface S is a surface for which the first variation of area vanishes.

This is equivalent to

$$\operatorname{tr}\theta = 0. \tag{197}$$

Then

$$\left.\frac{d^2 A}{dt^2}\right|_{t=0} = \int_S f \frac{\partial \operatorname{tr}\theta}{\partial t} \, d\mu_{\bar g}. \tag{198}$$

For a surface of *least area* one must have

$$\left.\frac{d^2 A}{dt^2}\right|_{t=0} \geq 0.$$

To obtain a suitable expression for $\frac{\partial \operatorname{tr}\theta}{\partial t}$ we consider the 2nd variation equations

$$\frac{\partial \theta_{ij}}{\partial t} = f \, \theta_{im}\theta^m_j - \bar\nabla_i \bar\nabla_j f - f \, R_{iNjN}.$$

(Here, we have an arbitrary local frame (E_i) for S, complemented with N.) We have

$$\begin{aligned}
\operatorname{tr}\theta &= \bar g^{ij}\theta_{ij}, \\
\frac{\partial \operatorname{tr}\theta}{\partial t} &= -\bar g^{im}\bar g^{jn} \frac{\partial \bar g_{mn}}{\partial t}\, \theta_{ij} + \bar g^{ij} \frac{\partial \theta_{ij}}{\partial t} \\
&= -2 f \, \theta^{ij}\theta_{ij} + \bar g^{ij} \frac{\partial \theta_{ij}}{\partial t}, \\
\bar g^{ij} \, R_{iNjN} &= \operatorname{Ric}(N,N) \quad (= R_{ij}N^i N^j).
\end{aligned}$$

Thus, taking the trace of the 2nd variation equations we obtain

$$\frac{\partial \operatorname{tr}\theta}{\partial t} = -\bar\Delta f - f \, (|\theta|^2 + \operatorname{Ric}(N,N)).$$

We now substitute for $\operatorname{Ric}(N,N)$ from the (twice contracted) Gauss equation (note that $\bar R_{ij} = K\bar g_{ij}$, $\bar R = 2K$, with K the Gauss curvature of S)

$$2K - (\operatorname{tr}\theta)^2 + |\theta|^2 = R - 2\operatorname{Ric}(N,N)$$

to obtain

$$\frac{\partial \operatorname{tr}\theta}{\partial t} = -\bar{\Delta} f + f\Big(K - \frac{1}{2}(\operatorname{tr}\theta)^2 - \frac{1}{2}|\theta|^2 - \frac{1}{2}R\Big).$$

Substituting in (198) and taking into account the fact that $\operatorname{tr}\theta = 0$ at $t = 0$ yields

$$\frac{d^2 A}{dt^2}\Big|_{t=0} = \int_S \Big\{ - f\,\bar{\Delta} f + f^2\Big(K - \frac{1}{2}|\theta|^2 - \frac{1}{2}R\Big)\Big\}\, d\mu_{\bar{g}}.$$

This must be ≥ 0 (surface of least area). Integrating by parts in the first term, this condition reads

$$\int_S K\, f^2\, d\mu_{\bar{g}} \geq \int_S \frac{1}{2}(|\theta|^2 + R)\, f^2\, d\mu_{\bar{g}} - \int_S |\bar{\nabla} f|^2 d\mu_{\bar{g}}. \tag{199}$$

Let Q_ρ be the closure of the interior of the coordinate cylinder C_ρ,

$$Q_\rho = \{(x^1, x^2, x^3) : (x^1)^2 + (x^2)^2 \leq \rho^2\}.$$

Setting

$$f = \begin{cases} 1 & \text{on } S \cap Q_\rho, \\ \frac{\log(\frac{\rho^2}{r})}{\log \rho} & \text{on } S \cap (Q_{\rho^2} \setminus Q_\rho), \\ 0 & \text{on } S \cap Q^c_{\rho^2}, \end{cases}$$

we have

$$\begin{aligned} \int_S |\bar{\nabla} f|^2\, d\mu_{\bar{g}} &= \int_{S\cap(Q_{\rho^2}\setminus Q_\rho)} |\bar{\nabla} f|^2\, d\mu_{\bar{g}} \\ &\leq C \int_\rho^{\rho^2} \left[\frac{d}{dr}\left(\frac{\log(\frac{\rho^2}{r})}{\log\rho}\right)\right]^2 r\, dr \\ &= \frac{C}{(\log\rho)^2}\int_\rho^{\rho^2}\frac{dr}{r} \\ &= \frac{C}{\log\rho} \to 0 \quad \text{as } \rho \to \infty. \end{aligned}$$

Taking then in (199) the limit $\rho \to \infty$ yields

$$\int_S K\, d\mu_{\bar{g}} \geq \int_S \frac{1}{2}(|\theta|^2 + R)\, d\mu_{\bar{g}} > 0. \tag{200}$$

(Recall that by Step 0 we can assume that $R > 0$ everywhere.) It follows that S is homeomorphic to a disk. (Recall the Cohn-Vossen inequality: $\int_S K\, d\mu_{\bar{g}} \leq 2\pi\,\chi(S)$, where $\chi(S)$ is the Euler characteristic of S.) We then consider the disks $D_\rho = S \cap Q_\rho$ and apply the Gauss–Bonnet theorem with boundary:

$$\int_{D_\rho} K \, d\mu_{\tilde{g}} = 2\pi - \int_{\partial D_\rho} \kappa \, ds,$$

where κ is the geodesic curvature of ∂D_ρ. The argument proceeds by showing that

$$\limsup_{\rho \to \infty} \int_{\partial D_\rho} \kappa \, ds \geq 2\pi.$$

Taking a sequence $\rho_i \to \infty$, achieving the lim sup then yields

$$\lim_{i \to \infty} \int_{D_{\rho_i}} K \, d\mu_{\tilde{g}} \leq 0$$

in *contradiction* with (200). This establishes Part 1.

Proof of Part 2. $M = 0$ implies that g is flat. The proof has two steps.

Step 0: If $M = 0$ and R does not vanish identically, then there exists a metric $\tilde{g}$ in the conformal class of g such that $\widetilde{M} < 0$ (N. O'Murchadha and J. York [21]).

By Part 1 we can then conclude that $M = 0$ implies that R vanishes identically.

Proof of Step 0. Setting $\tilde{g} = \Phi^4 g$ we solve the equation $\widetilde{R} = 0$ for Φ (see (188)):

$$-8 \, \triangle_g \, \Phi + R \, \Phi = \widetilde{R} \, \Phi^5 = 0.$$

We solve this linear equation under the asymptotic condition $\Phi \to 1$ at infinity. Since $R \geq 0$ the maximum principle applies and we obtain a function Φ which is everywhere positive (and less than or equal to 1). Since $M = 0$ the mass $\widetilde{M}$ of $\tilde{g}$ is contained in the function Φ. In fact we have

$$\begin{aligned} \widetilde{M} &= 2 \lim_{r \to \infty} r \, (\Phi - 1) \\ &= -\frac{1}{2\pi} \int_{\overline{M}} \triangle \, \Phi \, d\mu_g \\ &= -\frac{1}{16\pi} \int_{\overline{M}} R \, \Phi \, d\mu_g < 0. \end{aligned}$$

This proves Step 0.

Step 1: We are now given a metric g with $R = 0$ and $M = 0$. Consider the following variation of g (which may not be admissible because it may not satisfy the condition that the scalar curvature remains ≥ 0):

$$g_t = g + t \, \mathrm{Ric}(g),$$

with $t \in (-\epsilon, \epsilon)$ where ϵ is to be chosen suitably small. In general,

$$\frac{d\ R(g_t)}{dt}\bigg|_{t=0} = \dot{R}$$

is given in terms of $\dot{g} = \frac{dg_t}{dt}\big|_{t=0}$ by

$$\dot{R} = \nabla^i \nabla^j \dot{g}_{ij} - \triangle \operatorname{tr} \dot{g} - \dot{g}_{ij} R^{ij}. \tag{201}$$

(Here, ∇ is the covariant derivative with respect to $g_0 = g$.) The derivation of this formula is the following. Since $R = g^{ij} R_{ij}$ we have

$$\dot{R} = \underbrace{-g^{im} g^{jn} \dot{g}_{mn} R_{ij}}_{-\dot{g}_{ij} R^{ij}} + g^{ij} \dot{R}_{ij}.$$

Moreover,

$$\dot{R}_{ij} = \nabla_m \dot{\Gamma}^m_{ij} - \nabla_i \dot{\Gamma}^m_{mj},$$

where $\dot{\Gamma}^m_{ij}$ is the corresponding variation of the connection, the tensor field

$$\dot{\Gamma}^m_{ij} = \frac{1}{2}\, g^{mn}\, (\nabla_i \dot{g}_{jn} + \nabla_j \dot{g}_{in} - \nabla_n \dot{g}_{ij}).$$

Substituting in $\dot{R}_{ij}$ and taking the trace $g^{ij} \dot{R}_{ij}$ the result (201) follows.

Consider now the formula for $\dot{R}$ in the case of the variation $\dot{g}_{ij} = R_{ij}$. Then $\operatorname{tr} \dot{g} = R = 0$ and $\nabla^j \dot{g}_{ij} = \nabla^j R_{ij} = 0$ which is the twice contracted Bianchi identity in the case $R = 0$. Then (201) reduces to

$$\dot{R} = -|\operatorname{Ric}|^2.$$

Next, we correct the family g_t by a suitable conformal change to obtain an admissible family $\tilde{g}_t$. In fact we require $\widetilde{R}_t = 0$. Thus with $\tilde{g}_t = \Phi_t^4 g_t$ we stipulate (see (188))

$$\triangle_{g_t} \Phi_t - \frac{1}{8} R_t\, \Phi_t = 0, \qquad \Phi_t \to 1 \text{ at } \infty.$$

Here the maximum principle does not apply as R_t may not be ≥ 0 but since $R_0 = R = 0$, if ϵ is suitably small the problem can be solved by appealing to the implicit function theorem and yields $\Phi_t > 0$ everywhere. Now, since $R_{ij} = O(r^{-3})$ the mass of g_t is that of g, namely 0. Then the mass of $\tilde{g}_t$ is contained in the function Φ_t and is given by

$$\widetilde{M}_t = -\frac{1}{16\pi} \int_{\overline{M}} R_t\, \Phi_t\, d\mu_{g_t}. \tag{202}$$

Now for each $t \in (-\epsilon, \epsilon)$, ϵ suitably small, the metric $\tilde{g}_t$ defined in this way is admissible, that is, it satisfies the hypotheses of the theorem. Thus the result of Part 1 applies and we obtain $\widetilde{M}_t \geq 0$. Moreover, since $\widetilde{M}_0 = M = 0$, a minimum is achieved at $t = 0$. We thus have

$$\frac{d\widetilde{M}_t}{dt}\bigg|_{t=0} = 0.$$

Since $R_0 = 0$ and $\Phi_0 = 1$, the formula (202) gives

$$\frac{d\widetilde{M}_t}{dt}\bigg|_{t=0} = -\frac{1}{16\pi}\int_{\overline{M}} \dot{R}\, d\mu_{\bar{g}} = \frac{1}{16\pi}\int_{\overline{M}} |\operatorname{Ric}|^2\, d\mu_{\bar{g}}.$$

The vanishing of this implies $\operatorname{Ric}(g) = 0$ hence g is flat, $\overline{M}$ being 3-dimensional. This establishes Part 2. And with this the whole proof is completed. □

3.3 Angular momentum

Let us recall (see (149), (151), (158)) that the asymptotic quantity associated to the Killing field X of the background Minkowski metric is given by

$$Q_X = \lim_{r\to\infty} \int_{S_r} G_X^{*0i}\, dS_i. \tag{203}$$

The part of G_X^{*0i} which involves the first derivatives of X vanishes for the angular momentum (but does not vanish for the center of mass integrals (see below)). Taking X to be the generator of space rotations about the x^a coordinate axis, we obtain the angular momentum component

$$J_a = \lim_{r\to\infty} \int_{S_r} \underbrace{\epsilon_{abj}\, x^b\, p^i_j}_{=:A^i_a}\, dS_i, \tag{204}$$

where $p^i_j = k^i_j - \delta^i_j \operatorname{tr} k$, $k^i_j = \bar{g}^{im} k_{mj}$, and ϵ_{abc} is the fully antisymmetric 3-dimensional symbol.

Let us now assume that we have a complete spacelike hypersurface $\mathcal{H}$ which is strongly asymptotically flat. More precisely, let us assume that in an admissible chart in the neighbourhood of infinity in $\mathcal{H}$ we have

$$\bar{g}_{ij} = \left(1 + \frac{2M}{r}\right)\delta_{ij} + O_2(r^{-1-\epsilon}), \tag{205}$$

$$k_{ij} = O_1(r^{-2-\epsilon}) \tag{206}$$

for some $\epsilon > 0$. This implies that $P_j = 0$.

3.3.1 Independence from exhaustion. Let $\mathcal{U}_2 \supset \mathcal{U}_1$ be domains in $\mathcal{H}$ with $\mathcal{U}_1 \supset B_R$ where B_R is the coordinate ball of radius R. Then we have

$$\int_{\partial\mathcal{U}_2} A_a^i \, dS_i - \int_{\partial\mathcal{U}_1} A_a^i \, dS_i = \int_{\mathcal{U}_2\setminus\mathcal{U}_1} \partial_i \, A_a^i \, d^3x.$$

Writing

$$\partial_i A_a^i = \epsilon_{aij} \, p_j^i + \epsilon_{abj} \, x^b \, \partial_i p_j^i,$$

the first term is

$$\frac{1}{2} \, \epsilon_{aij} \, (p_j^i - p_i^j)$$

and the difference $p_j^i - p_i^j$ reads $p_j^i - p_i^j = (\bar{g}^{im} - \delta^{im})k_{mj} - (\bar{g}^{jm} - \delta^{jm})k_{mi}$. Thus, $\frac{1}{2} \, \epsilon_{aij} \, (p_j^i - p_i^j) = O(r^{-3-\epsilon})$. In regard to the second term we express

$$\partial_i p_j^i = \nabla_i p_j^i - \Gamma_{im}^i p_j^m + \Gamma_{ij}^m p_m^i.$$

Since by the Codazzi constraint $\nabla_i p_j^i = 0$, we then obtain $\partial_i p_j^i = O(r^{-4-\epsilon})$. Thus, the second term is $O(r^{-3-\epsilon})$ as well. Therefore,

$$\left| \int_{\mathcal{U}_2\setminus\mathcal{U}_1} \partial_i \, A_a^i \, d^3x \right| \leq C \int_{\mathcal{U}_2\setminus\mathcal{U}_1} r^{-3-\epsilon} \, d^3x \leq C' R^{-\epsilon} \to 0. \tag{207}$$

3.3.2 Conservation of angular momentum. Consider the following limit on each level hypersurface of the canonical maximal time function:

$$\lim_{r\to\infty} r^2 \, \frac{\partial A_a^i(r\xi)}{\partial t} = B_a^i(\xi), \tag{208}$$

for each $\xi \in S^2 \subset \mathbb{R}^3$ (r, ξ are polar coordinates in $\mathbb{R}^3$). Then we have

$$\frac{\partial J_a}{\partial t} = \int_{\xi\in S^2} B_a^i(\xi) \, \xi^i \, d\mu_\xi \tag{209}$$

where $d\mu_\xi$ is the standard measure on $S^2 = \{|\xi| = 1\} \subset \mathbb{R}^3$. Since $\operatorname{tr} k = 0$, A_a^i reduces to

$$A_a^i = -\frac{1}{2} \, \epsilon_{abj} \, x^b \, \sqrt{\det \bar{g}} \, k_j^i. \tag{210}$$

Taking account of the fact that

$$\frac{\partial \sqrt{\det \bar{g}}}{\partial t} = \sqrt{\det \bar{g}} \, \Phi \operatorname{tr} k = 0,$$

we then have

$$\frac{\partial A_a^i}{\partial t} = -\frac{1}{2} \, \epsilon_{abj} \, x^b \, \sqrt{\det \bar{g}} \, \frac{\partial k_j^i}{\partial t}. \tag{211}$$

Substituting from (178), (179) this becomes

$$\frac{\partial A^i_a}{\partial t} = -\frac{1}{2}\,\epsilon_{abj}\,x^b\,\sqrt{\det \bar{g}}\,(\bar{\nabla}^i\bar{\nabla}_j\Phi - \bar{R}^i_j\Phi). \tag{212}$$

Now, only the part of $\bar{\nabla}^i\bar{\nabla}_j\Phi - \bar{R}^i_j\Phi$ which is $O(r^{-3})$ contributes to the limit B^i_a. Under the assumptions (205), (206), the lapse equation (180) implies

$$\Phi = 1 - \frac{N}{r} + O_2(r^{-1-\epsilon}),$$

where $N = \frac{1}{4\pi}\int_{\mathcal{H}} \bar{\Delta}\,\Phi\,d\mu_{\bar{g}} = \frac{1}{4\pi}\int_{\mathcal{H}} |k|^2\Phi\,d\mu_{\bar{g}}$. Hence

$$\bar{\nabla}_i\,\Phi = \frac{N}{r^2}\,\xi^i + O_1(r^{-2-\epsilon}),$$

and

$$\bar{\nabla}^i\,\bar{\nabla}_j\Phi = \frac{(\delta^i_j - 3\xi^i\xi_j)\,N}{r^3} + O(r^{-3-\epsilon}).$$

Also (205) implies

$$\bar{R}^i_j = \frac{M}{r^3}\,(\delta^i_j - 3\xi^i\xi_j) + O(r^{-3-\epsilon}).$$

It follows that $B^i_a(\xi)$ is given by

$$\begin{aligned} B^i_a\,(\xi) &= -\frac{1}{2}\,\epsilon_{abj}\,\xi^b\,(N-M)\,(\delta^i_j - 3\xi^i\xi_j) \\ &= -\frac{1}{2}\,(N-M)\,\epsilon_{abi}\,\xi^b, \end{aligned}$$

which implies

$$B^i_a\,(\xi)\,\xi^i = 0. \tag{213}$$

In view of the formula (209) the conservation of angular momentum follows.

3.4 The center of mass integrals

The center of mass integrals C_j correspond to the vector fields

$$X_{(j)} = x^j\,\frac{\partial}{\partial t} + t\,\frac{\partial}{\partial x^j} \tag{214}$$

which generate Lorentz boosts. We have (see (149), (160), (163))

$$
\begin{aligned}
G_X^{*0i} &= F_0^{*0i}\, x^j + t\, F_j^{*0i} + K_X^{*0i} \qquad (215)\\
&= -e^i\, x^j + t\, p^i_j + \underbrace{\frac{1}{4}\{\, (1-\Phi^{-1}\sqrt{\det \bar g})\,\delta^i_j + (\delta^i_j - \Phi\,\sqrt{\det \bar g}\,\bar g^{ij})\,\}}_{=:\, q^i_j}.
\end{aligned}
$$

The center of mass integrals C_j are then given by

$$
C_j = -\lim_{r\to\infty}\int_{S_r} G_X^{*0i}\, dS_i = \lim_{r\to\infty}\int_{S_r} (e^i\, x^j - t\, p^i_j - q^i_j)\, dS_i . \qquad (216)
$$

Remark. Note that

$$
\lim_{r\to\infty}\int_{S_r} p^i_j\, dS_i = P_j
$$

and $P_j = 0$ by (206). Thus, (216) simplifies to

$$
C_j = \lim_{r\to\infty}\int_{S_r} (e^i\, x^j - q^i_j)\, dS_i . \qquad (217)
$$

We are going to present a sketch of the proof that the limit (217) exists for an exhaustion by coordinate spheres.

Sketch of the proof. Given $r_2 > r_1$ we express

$$
\begin{aligned}
&\int_{S_{r_2}} (e^i\, x^j - q^i_j) dS_i - \int_{S_{r_1}} (e^i\, x^j - q^i_j) dS_i \\
&= \int_{B_{r_2}\setminus B_{r_1}} \underbrace{\partial_i\, (e^i\, x^j - q^i_j)}_{=:\omega_j}\, d^3x \\
&= \underbrace{\int_{r_1}^{r_2} \left\{ \int_{\xi\in S^2} \omega_j(r\xi)\, d\mu_\xi \right\} r^2\, dr}_{=:D(r_1,r_2)} .
\end{aligned}
$$

The hypotheses (205) and (206) imply that

$$
\omega_j\,(r\xi) = \frac{K\,\xi_j}{r^3} + O(r^{-3-\epsilon})
$$

where K is a quadratic expression in M, N, and thus depends only on t. Since

$$
\int_{\xi\in S^2} \xi_j\, d\mu_\xi = 0,
$$

we then obtain

$$\int_{\xi \in S^2} \omega_j(r\xi)\, d\mu_\xi = O(r^{-3-\epsilon}).$$

It follows that

$$\begin{aligned} |D(r_1, r_2)| &\leq C \int_{r_1}^{r_2} r^{-1-\epsilon}\, dr \\ &\leq C'\, r_1^{-\epsilon} \to 0, \quad \text{as } r_1 \to \infty. \end{aligned}$$

This proves that the limit (217) exists.

We now investigate the *gauge invariance* of the C_j; that is, the invariance of the C_j under spatial gauge transformations. *Recall that thus we are subjecting the metric g to a diffeomorphism but we are not subjecting the background Minkowski structure to any transformation.* Moreover, we are only allowing transformations which do not change the asymptotic form (205) and (206).

However, we do not expect the C_j to be invariant under those transformations which are asymptotic to non-trivial space translations at infinity. For, in Classical Mechanics the center of mass integrals are affected by space translations. We thus restrict ourselves to transformations which are asymptotic to the identity at infinity. Such transformations are generated by vector fields Y whose components Y^i in an admissible coordinate system in the neighbourhood of infinity satisfy

$$Y^i = O_3(r^{-\epsilon}).$$

The variations of Φ and $\bar{g}_{ij}$ are

$$\dot{\Phi} = \mathcal{L}_Y \Phi = Y^i\, \partial_i\, \Phi = O_1(r^{-2-\epsilon}), \tag{218}$$

$$\begin{aligned} \dot{\bar{g}}_{ij} &= \mathcal{L}_Y \bar{g}_{ij} = Y^k\, \partial_k\, \bar{g}_{ij} + \bar{g}_{kj}\, \partial_i\, Y^k + \bar{g}_{ik}\, \partial_j\, Y^k \\ &= \underbrace{\partial_i\, Y^j + \partial_j\, Y^i}_{=O_2(r^{-1-\epsilon})} + O_1(r^{-2-\epsilon}). \end{aligned} \tag{219}$$

We then obtain

$$\begin{aligned} \dot{e}^i &= \frac{1}{4}\, (\delta^{im}\, \delta^{jn} - \delta^{ij}\, \delta^{mn})\, \partial_j\, (\partial_m\, Y^n + \partial_n\, Y^m) + O(r^{-3-\epsilon}) \\ &= \frac{1}{4}\, \partial_k\, (\partial_k\, Y^i - \partial_i\, Y^k) + O(r^{-3-\epsilon}). \end{aligned} \tag{220}$$

Furthermore we have

$$\dot{q}^i_j = \frac{1}{4}\, \{ -(\sqrt{\det \bar{g}})^{\cdot}\, \delta^i_j - (\sqrt{\det \bar{g}}\, \bar{g}^{ij})^{\cdot} \} + O(r^{-2-\epsilon})$$

and

$$(\sqrt{\det \bar{g}})^{\cdot} = \partial_k Y^k + O(r^{-2-\epsilon}),$$
$$(\sqrt{\det \bar{g}}\, \bar{g}^{ij})^{\cdot} = \delta^i_j\, \partial_k Y^k - \partial_i Y^j - \partial_j Y^i + O(r^{-2-\epsilon}).$$

Therefore we obtain

$$\dot{q}^i_j = -\frac{1}{4}(2\,\delta^i_j\, \partial_k Y^k - \partial_i Y^j - \partial_j Y^i) + O(r^{-2-\epsilon}). \tag{221}$$

The terms of order $O(r^{-3-\epsilon})$ in (220) and $O(r^{-2-\epsilon})$ in (221) do not contribute to $\dot{C}_j$. Thus, one finds

$$\begin{aligned}\dot{C}_j &= \lim_{r\to\infty} \frac{1}{4} \int_{S_r} \{\partial_k (\partial_k Y^i - \partial_i Y^k)\, x^j \\ &\quad + 2\delta^i_j\, \partial_k Y^k - \partial_i Y^j - \partial_j Y^i\}\, dS_i \end{aligned} \tag{222}$$

$$\begin{aligned} &= \lim_{r\to\infty} \frac{1}{4} \int_{B_r} \partial_i \{\partial_k (\partial_k Y^i - \partial_i Y^k) x^j \\ &\quad + 2\,\delta^i_j\, \partial_k Y^k - \partial_i Y^j - \partial_j Y^i\}\, d^3x. \end{aligned} \tag{223}$$

Now $\partial_i \partial_k(\partial_k Y^i - \partial_i Y^k) = 0$ and $\partial_i x^j = \delta^i_j$. Therefore the integrant in the volume integrals is

$$\partial_k(\partial_k Y^j - \partial_j Y^k) + 2\partial_j \partial_k Y^k - \partial_i \partial_i Y^j - \partial_i\, \partial_j Y^i = 0, \tag{224}$$

which proves the gauge invariance of the C_j.

Remark. In the expression for C_j we can replace q^i_j by $\bar{q}^i_j$:

$$\bar{q}^i_j = \frac{1}{4}\{(1 - \sqrt{\det \bar{g}})\, \delta^i_j + (\delta^i_j - \sqrt{\det \bar{g}}\, \bar{g}^{ij})\}, \tag{225}$$

obtained by replacing Φ by 1 in the expression for q^i_j.

Proof of the remark. Under a variation $\delta\Phi$ of Φ we have

$$\begin{aligned}\delta\, C_j &= \lim_{r\to\infty} \int_{S_r} -\delta\, q^i_j\, dS_i \\ &= -\frac{1}{4} \lim_{r\to\infty} \int_{S_r} (\Phi^{-2}\, \delta^i_j - \bar{g}^{ij})\, \delta\, \Phi\, \sqrt{\det \bar{g}}\, dS_i. \end{aligned} \tag{226}$$

Now, with

$$\Phi = -\frac{N}{r} + O(r^{-1-\epsilon})$$

the following hold:

$$\Phi^{-2}\,\delta^i_j - \bar{g}^{ij} = 2\,\frac{N}{r}\,\delta^i_j + \frac{2M}{r}\,\delta^i_j + O(r^{-1-\epsilon}),$$
$$\delta\Phi = -\frac{\delta N}{r} + O(r^{-1-\epsilon}).$$

Consequently,

$$\delta C_j = \frac{1}{2}\lim_{r\to\infty}\int_{S_r}\frac{\overbrace{\delta N\,(N+M)}^{=:C}}{r^2}\,dS_j = \frac{C}{2}\int_{\xi\in S_2}\xi_j\,d\mu_\xi = 0$$

which proves the above remark.

We have thus reached the conclusion that the

$$C_j = \lim_{r\to\infty}\int_{S_r}(e^i\,x^j - \bar{q}^i_j)\,dS_i$$

are *invariant under gauge transformations which are asymptotic to the identity at infinity*. They correspond in Classical Mechanics to the components of the *moment* of a mass distribution about a given origin, the product of the total mass times the components of the position vector of the *center of mass* relative to that origin.

3.4.1 Conservation of center of mass integrals. Consider

$$\frac{d\,C_j}{d\,t} = \int_{\xi\in S^2} D_j\,(\xi)\,d\mu_\xi, \tag{227}$$

where

$$D_j(\xi) = \lim_{r\to\infty} r^2\Big(\frac{\partial e^i}{\partial t}x^j - \frac{\partial \bar{q}^i_j}{\partial t}\Big)(r\xi).$$

From (225) we obtain

$$\frac{\partial \bar{q}^i_j}{\partial t} = -\frac{1}{4}\,\sqrt{\det\bar{g}}\,\frac{\partial \bar{g}^i_j}{\partial t} = \frac{1}{2}\,\sqrt{\det\bar{g}}\,\Phi\,k^{ij} = O(r^{-2-\epsilon}).$$

(Recall that for a maximal time function we have $\frac{\partial}{\partial t}\sqrt{\det\bar{g}} = 0$.) Similarly,

$$\frac{\partial e^i}{\partial t} = \sqrt{\det\bar{g}}\,\frac{1}{4}\,(\bar{g}^{in}\,\bar{g}^{jm} - \bar{g}^{ij}\,\bar{g}^{mn})\,\partial_j\,\underbrace{(\partial_t\,\bar{g}_{mn})}_{=2\Phi k_{mn}} + O(r^{-4-\epsilon})$$

and

$$\begin{aligned}\partial_j\,(\Phi\,k_{mn}) &= \nabla_j\,(\Phi\,k_{mn}) + \Gamma^l_{jm}\,\Phi\,k_{ln} + \Gamma^l_{jn}\,\Phi\,k_{ml}\\ &= \nabla_j\,(\Phi\,k_{mn}) + \;O(r^{-4-\epsilon})\\ &= \Phi\,\nabla_j\,k_{mn} + O(r^{-4-\epsilon}).\end{aligned}$$

Hence we obtain

$$\begin{aligned}\frac{\partial e^i}{\partial t} &= \frac{1}{2}\,\sqrt{\det\bar{g}}\;\Phi\,(\nabla_j\,k^{ij} - \bar{g}^{ij}\,\nabla_j\,\mathrm{tr}\,k) + O(r^{-4-\epsilon})\\ &= O(r^{-4-\epsilon}),\end{aligned}$$

the first term vanishing by virtue of the Codazzi equations.

The above imply that

$$D_j\,(\xi) = \lim_{r\to\infty} r^2\left(\frac{\partial e^i}{\partial t}\,x^j - \frac{\partial \bar{q}^i_j}{\partial t}\right)(r\,\xi) = 0. \tag{228}$$

We conclude that

$$\frac{\partial\,C_j}{\partial\,t} = 0. \tag{229}$$

We have thus established the conservation of the center of mass integrals.

4 The global stability of Minkowski spacetime

In this chapter we shall first state the problem of the global stability of Minkowski spacetime. We shall then treat simpler analogous problems arising in field theories in a given spacetime. Finally we shall conclude the volume with a sketch of the proof of the global stability theorem of Minkowski spacetime.

4.1 Statement of the problem

The *Minkowski spacetime* is the simplest solution of the Einstein equations. This is the spacetime manifold of Special Relativity,

$$(\mathbb{R}^4, \eta), \quad \text{where } \eta_{\mu\nu} = \text{diag}(-1, 1, 1, 1) \text{ in rectangular coordinates.}$$

This is *geodesically complete*: Every geodesic can be continued ad infinitum in affine parameter.

The function x^0 corresponding to any rectangular coordinate system is a canonical maximal time function. The level sets Σ_t ($x^0 = t$) are maximal spacelike hypersurfaces. Here they happen to be totally geodesic (not only is $\operatorname{tr} k = 0$ but $k_{ij} = 0$ identically). They are also globally parallel, that is the lapse function is $\Phi = 1$ identically. (Recall the equation $\bar{\triangle}\Phi - |k|^2\Phi = 0$.)

Cauchy problem with initial data on a complete asymptotically flat maximal hypersurface. Consider initial data sets $(\mathcal{H}_0, \bar{g}_0, k_0)$ with $\mathcal{H}_0$ diffeomorphic to $\mathbb{R}^3$ and $\operatorname{tr} k_0 = 0$, which are strongly asymptotically flat. That is, there exists a coordinate system in the neighbourhood of infinity in which the metric coefficients obey

$$\bar{g}_{0_{ij}} = \left(1 + \frac{2M}{r}\right)\delta_{ij} + O_2\,(r^{-1-\epsilon}), \quad \epsilon > 0, \tag{230}$$

and we have

$$k_0 = O_1\,(r^{-2-\epsilon}). \tag{231}$$

Note that the total linear momentum of such an initial data set vanishes: $P^i = 0$.

The initial data set is, moreover, required to satisfy the constraint equations

$$\text{Codazzi equations:} \quad \bar{\nabla}^i k_{ij} = 0, \tag{232}$$

$$\text{Gauss equation:} \quad \bar{R} = |k|^2. \tag{233}$$

Our hypotheses are that we are given such an initial data set. Now the problem is the following:

Problem. Supplement these hypotheses by a suitable smallness condition and show that we can then construct a geodesically complete solution of the Einstein equations, tending to the Minkowski spacetime along any geodesic.

4.1.1 Field theories in a given spacetime. Consider Lagrangians of the form

$$L = L^* \, d\mu_g,$$

where L^* is a scalar function, constructed out of the fields, their exterior or more generally covariant derivatives, the metric and connection coefficients. We give the following three examples.

Example 1 (Scalar field Φ). Let

$$\begin{aligned} \sigma &= g^{\mu\nu} \, \partial_\mu \, \Phi \, \partial_\nu \, \Phi, \\ L^* &= L^* \, (\sigma). \end{aligned}$$

Here only exterior derivatives are involved. So, the Lagrangian does not depend on the connection coefficients.

Example 2 (Electromagnetic field $F_{\mu\nu}$). Here the spacetime manifold is 4-dimensional and we have

$$\begin{aligned} F &= dA & &(F_{\mu\nu} = \partial_\mu \, A_\nu - \partial_\nu \, A_\mu), \\ \alpha &= F^{\mu\nu} \, F_{\mu\nu}; & &F^{\mu\nu} = g^{\mu\kappa} \, g^{\nu\lambda} \, F_{\kappa\lambda}, \\ \beta &= F^{\mu\nu} \, {}^*F_{\mu\nu}; & &{}^*F_{\mu\nu} = \frac{1}{2} \, F^{\kappa\lambda} \, \epsilon_{\kappa\lambda\mu\nu}, \\ L^* &= L^* \, (\alpha, \beta). \end{aligned}$$

Also here, only exterior derivatives are involved and the Lagrangian does not depend on the connection coefficients.

Example 3 (Problem in Riemannian Geometry). Let (M, g) be a compact Riemannian manifold. Find a vector field U generating an approximate isometry of (M, g).

The solution is the following: Consider

$$\pi = \mathcal{L}_U \, g,$$

the deformation tensor corresponding to U. This measures the deviation of the 1-parameter group of diffeomorphisms generated by U from a group of isometries. We then minimize

$$\int_M |\pi|^2 \, d\mu_g$$

under the constraint

$$\int_M |U|^2 \, d\mu_g = 1.$$

The Euler–Lagrange equation is the eigenvalue problem

$$\operatorname{div} \pi + \lambda \, U = 0,$$

that is

$$\nabla_j \, \pi^{ij} + \lambda \, U^i = 0,$$

where λ is the Lagrange multiplier or eigenvalue. For the analogous problem on a spacetime manifold

$$\begin{aligned} L^* \, d\mu_g &= \pi^{\mu\nu} \, \pi_{\mu\nu} \, d\mu_g, \\ \pi^{\mu\nu} &= g^{\mu\nu} \, g^{\nu\lambda} \, \pi_{\kappa\lambda}, \\ \pi_{\mu\nu} &= \mathcal{L}_U g_{\mu\nu} = \nabla_\mu \, U_\nu + \nabla_\nu \, U_\mu, \\ U_\mu &= g_{\mu\nu} \, U^\nu. \end{aligned}$$

Here L^* depends on the covariant derivatives of U, therefore the connection coefficients corresponding to g.

This problem has a direct application to a problem in General Relativity. The problem in General Relativity is that of preservation of symmetry. Namely, to show that if the initial conditions possess a continuous isometry group, then the solution also possesses the same isometry group. The difficulty in General Relativity, which is absent for the analogous problem in the case of a theory in a given spacetime, is to extend the action of the group from the initial hypersurface to the spacetime manifold. More precisely, the problem can be formulated as follows. Given an initial data set $(\mathcal{H}, \bar{g}, k)$ for the Einstein equations for which there exists a vector field $\bar{U}$ on $\mathcal{H}$ such that $\mathcal{L}_{\bar{U}} \bar{g} = \mathcal{L}_{\bar{U}} k = 0$, the problem is to extend $\bar{U}$ to a vector field U defined on the maximal development (M, g) so that U is a Killing field of g: $\mathcal{L}_U g = 0$. It turns out that a suitable way to define U is to require that it satisfies the Euler–Lagrange equation corresponding to the above Lagrangian, namely the equation

$$\operatorname{div} \pi = 0.$$

For a field theory in a given spacetime, the Lagrangian not depending on any other underlying structure on the manifold other than the metric and the corresponding connection coefficients, we can define the *gravitational stress* $T^{\mu\nu}$ by considering the response of the action

$$\mathcal{A}\,[\mathcal{U}] = \int_{\mathcal{U}} L\, d\mu_g$$

to variations of the underlying metric g. We define $T^{\mu\nu}$ by the condition that

$$\dot{\mathcal{A}} = -\frac{1}{2}\int_{\mathcal{U}} T^{\mu\nu}\, \dot{g}_{\mu\nu}\, d\mu_g$$

for all variations $\dot{g}_{\mu\nu}$ with support in $\mathcal{U}$, for any domain $\mathcal{U}$ with compact closure in M. By definition $T^{\mu\nu}$ is symmetric. Thus

$$T^{\nu\mu} = T^{\mu\nu}.$$

Proposition 5. *By virtue of the Euler–Lagrange equations for the matter fields, $T^{\mu\nu}$ is conserved. That is*

$$\nabla_\nu\, T^{\mu\nu} = 0.$$

Proof. Let us denote by Φ the collection of matter fields. Consider any smooth vector field X with compact support in $\mathcal{U}$. This generates a 1-parameter group $\{f_t\}$ of diffeomorphisms of $\mathcal{U}$ onto itself leaving the action invariant. That is, if we denote the action by $\mathcal{A}[g, \Phi; \mathcal{U}]$, then, replacing g by $g_t = f_t^{*}g$ and Φ by $\Phi_t = f_t^{*}\Phi$ (schematically; whatever the action of f_t on Φ is; for example, if Φ is a vector field, the action is f_{-t*}, the push-forward by $f_{-t} = f_t^{-1}$) we have

$$\mathcal{A}\,[g_t, \Phi_t; \mathcal{U}] = \mathcal{A}\,[g, \Phi; \mathcal{U}].$$

So $\mathcal{A}(t) := \mathcal{A}[g_t, \Phi_t; \mathcal{U}]$ is independent of t. We thus have

$$0 = \frac{d\mathcal{A}}{dt}\bigg|_{t=0} = \int_{\mathcal{U}} \left\{ -\frac{1}{2}\, T^{\mu\nu}\, \dot{g}_{\mu\nu} + E\, \dot{\Phi} \right\} d\mu_g = 0.$$

E is the variation of $\mathcal{A}$ with respect to Φ, which vanishes by virtue of the Euler–Lagrange equations. Moreover, one has

$$\dot{g} = \mathcal{L}_X g, \quad \dot{g}_{\mu\nu} = \nabla_\mu X_\nu + \nabla_\nu X_\mu.$$

Thus we find

$$\begin{aligned} 0 &= -\int_{\mathcal{U}} \frac{1}{2}\, T^{\mu\nu}\, (\nabla_\mu X_\nu + \nabla_\nu X_\mu)\, d\mu_g \\ &= -\int_{\mathcal{U}} T^{\mu\nu}\, \nabla_\nu X_\mu\, d\mu_g \\ &= \int_{\mathcal{U}} (\nabla_\nu T^{\mu\nu})\, X_\mu\, d\mu_g. \end{aligned}$$

As this holds for arbitrary smooth X of compact support in $\mathcal{U}$, the result follows. □

Let us write down the gravitational stress corresponding to the three examples above.

Example 1 (Scalar field). We have

$$\frac{\partial \sigma}{\partial g^{\mu\nu}} = \partial_\mu \, \Phi \, \partial_\nu \, \Phi.$$

Then

$$T^{\mu\nu} = 2 \, \frac{dL^*}{d\sigma} \, \partial^\mu \, \Phi \, \partial^\nu \, \Phi - L^* \, g^{\mu\nu}.$$

Example 2 (Electromagnetic field). We have

$$\frac{\partial \alpha}{\partial g^{\mu\nu}} = 2 \, F_{\mu\kappa} \, F_\nu{}^\kappa$$

and

$$\frac{\partial \beta}{\partial g^{\mu\nu}} = \frac{1}{2} \, \beta \, g_{\mu\nu}, \quad \text{in view of the identity } F_{\mu\kappa} \, {}^*F_\nu{}^\kappa = \frac{1}{4} \, \beta \, g_{\mu\nu}.$$

(It is straightforward to check this identity in an arbitrary orthonormal frame.) The gravitational stress is then given by

$$T^{\mu\nu} = 4 \, \frac{\partial L^*}{\partial \alpha} \, F^\mu_\kappa \, F^{\nu\kappa} + \left(\beta \, \frac{\partial L^*}{\partial \beta} - L^* \right) g^{\mu\nu}.$$

Example 3. In this case,

$$L^* = \pi^{\mu\nu} \, \pi_{\mu\nu}.$$

Under a variation of g,

$$\dot{\pi} = \mathcal{L}_U \dot{g}.$$

We then have

$$\dot{\mathcal{A}} = \int_{\mathcal{U}} \left\{ \left(-2 \, \pi^\mu_\kappa \, \pi^{\nu\kappa} + \frac{1}{2} \, \pi^{\kappa\lambda} \, \pi_{\kappa\lambda} \, g^{\mu\nu} \right) \dot{g}_{\mu\nu} + 2 \, \pi^{\mu\nu} \, \dot{\pi}_{\mu\nu} \right\} d\mu_g,$$

$$\dot{\pi}_{\mu\nu} = \mathcal{L}_U \, \dot{g}_{\mu\nu} = U^\kappa \, \nabla_\kappa \, \dot{g}_{\mu\nu} + \dot{g}_{\kappa\nu} \, \nabla_\mu \, U^\kappa + \dot{g}_{\kappa\mu} \, \nabla_\nu \, U^\kappa,$$

and, integrating by parts,

$$\int_{\mathcal{U}} \pi^{\mu\nu} \, \dot{\pi}_{\mu\nu} \, d\mu_g = \int_{\mathcal{U}} \{ -\nabla_\kappa \, (U^\kappa \, \pi^{\mu\nu}) + \pi^{\kappa\nu} \, \nabla_\kappa \, U^\mu + \pi^{\kappa\mu} \, \nabla_\kappa \, U^\nu \} \, \dot{g}_{\mu\nu} \, d\mu_g.$$

We thus find that

$$T^{\mu\nu} = 4 \, \pi^\mu_\kappa \, \pi^{\nu\kappa} - \pi^{\kappa\lambda} \, \pi_{\kappa\lambda} \, g^{\mu\nu} + 4 \, \{ \nabla_\kappa \, (U^\kappa \, \pi^{\mu\nu}) - \pi^{\kappa\nu} \, \nabla_\kappa \, U^\mu - \pi^{\kappa\mu} \, \nabla_\kappa \, U^\nu \}.$$

Let us now consider the conformal properties.

Proposition 6. *If the action in a domain $\mathcal{U}$ is invariant under conformal transformations of the metric which differ from the identity in a subdomain with compact closure in $\mathcal{U}$, then the gravitational stress is trace-free.*

For, with

$$\tilde{g} = \Omega^2 g,$$

where Ω differs from 1 in a domain with compact closure in $\mathcal{U}$, let

$$\mathcal{A}[\tilde{g}; \mathcal{U}] = \mathcal{A}[g; \mathcal{U}].$$

Then with

$$\dot{g} = \lambda g, \quad \text{where } \lambda = 2\,\Omega\,\dot{\Omega}$$

has compact support in $\mathcal{U}$, we have

$$\dot{\mathcal{A}} = 0.$$

Since

$$\dot{\mathcal{A}} = -\frac{1}{2}\int_{\mathcal{U}} T^{\mu\nu}\,\dot{g}_{\mu\nu}\,d\mu_g,$$

this reads

$$0 = -\frac{1}{2}\int_{\mathcal{U}} \lambda \operatorname{tr} T\,d\mu_g.$$

As this holds for an arbitrary function λ with compact support in $\mathcal{U}$, it follows that $\operatorname{tr} T = 0$.

Example (Maxwell Lagrangian for electromagnetic theory in a vacuum, $L^* = \frac{1}{4}\alpha$). Under the transformation $g \mapsto \Omega^2 g$ we have

$$\begin{aligned} F_{\mu\nu} &\mapsto F_{\mu\nu}, & {}^*F_{\mu\nu} &\mapsto {}^*F_{\mu\nu}, \\ L^* &\mapsto \Omega^{-4} L^*, & L = L^* d\mu_g &\mapsto L^* d\mu_g = L, \end{aligned}$$

the spacetime manifold being 4-dimensional. Thus the action is conformally invariant. Then

$$T^{\mu\nu} = F^{\mu}_{\kappa}\,F^{\nu\kappa} - \frac{1}{4}\,F^{\kappa\lambda}\,F_{\kappa\lambda}\,g^{\mu\nu}$$

and indeed we have

$$\operatorname{tr} T = 0.$$

Now let X be a vector field on M. Consider the vector field

$$P^\mu = -T^\mu_\nu\, X^\nu.$$

We have

$$\begin{aligned}\nabla\cdot P &= -(\nabla_\mu\, T^\mu_\nu)\, X^\nu - T^\mu_\nu\, \nabla_\mu\, X^\nu \\ &= -\frac{1}{2}\, T^{\mu\nu}\, (\nabla_\mu\, X_\nu + \nabla_\nu\, X_\mu),\end{aligned}$$

by virtue of the symmetry of $T^{\mu\nu}$. That is,

$$\nabla\cdot P = -\frac{1}{2}\, T^{\mu\nu}\, \pi_{\mu\nu},$$

where $\pi = \mathcal{L}_X g$. Moreover, if $T^{\mu\nu}$ is trace-free, we can replace $\pi_{\mu\nu}$ by its trace-free part $\hat{\pi}_{\mu\nu}$ in the above formula. In the case that $\dim M = 4$ this trace-free part is

$$\hat{\pi}_{\mu\nu} = \pi_{\mu\nu} - \frac{1}{4}\, g_{\mu\nu}\, \mathrm{tr}\, \pi.$$

We conclude that – in general – if X generates a 1-parameter group of isometries of (M, g) (that is, X is a Killing field), then P is divergence-free:

$$\nabla\cdot P = 0.$$

If $\mathcal{A}$ is conformally invariant, the same holds if X generates a 1-parameter group of conformal isometries of (M, g) (that is, X is a conformal Killing field). Let us recall here that *a diffeomorphism f of M is called an isometry if $f^*g = g$*. It is called a conformal isometry if there exists a positive function Ω such that $f^*g = \Omega^2 g$.

In the case that the canonical stress is related to the gravitational stress by

$$T^{*\mu}_{\nu} = T^{\mu\lambda} g_{\lambda\nu}, \tag{234}$$

the conservation of P is equivalent to Noether's theorem, since P then coincides with the Noether current.

The divergence theorem gives an integral conservation law corresponding to the differential law

$$\nabla\cdot P = 0.$$

We consider the dual 3-form ${}^*P$, writing

$${}^*P_{\alpha\beta\gamma} = P^\mu\, \epsilon_{\mu\alpha\beta\gamma}. \tag{235}$$

Then it holds that

$$\nabla\cdot P = 0 \iff d\, {}^*P = 0. \tag{236}$$

That is, P being divergence-free is equivalent to $^*P$ being closed.

Let H_1 and H_2 be homologous hypersurfaces in M. If H_1 and H_2 have boundary, then $\partial H_1 = \partial H_2$. We can then apply the divergence theorem to the domain bounded by H_1 and H_2 to obtain

$$\int_{H_2} {}^*P = \int_{H_1} {}^*P.$$

We can also apply the theorem to two complete Cauchy hypersurfaces H_1 and H_2 as in Figure 12.

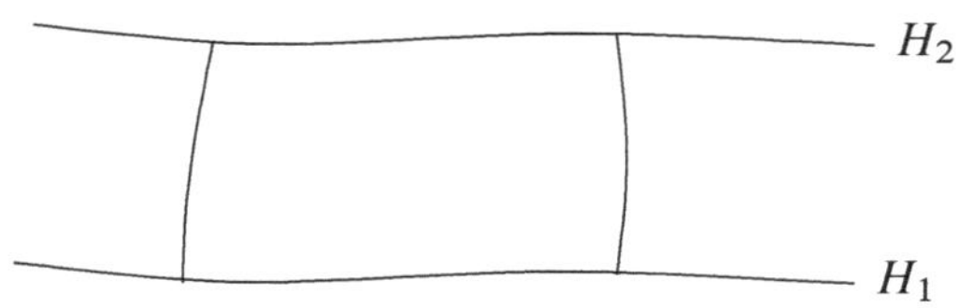

Figure 12

If we can show that the lateral contribution tends to 0, we again obtain the conservation law

$$\int_{H_2} {}^*P = \int_{H_1} {}^*P.$$

Otherwise, we can take the lateral hypersurface to be an incoming null hypersurface as in Figure 13, in which case its contribution will be non-negative by virtue of the physical requirement which follows. We then obtain the inequality

$$\int_{H_2} {}^*P \leq \int_{H_1} {}^*P.$$

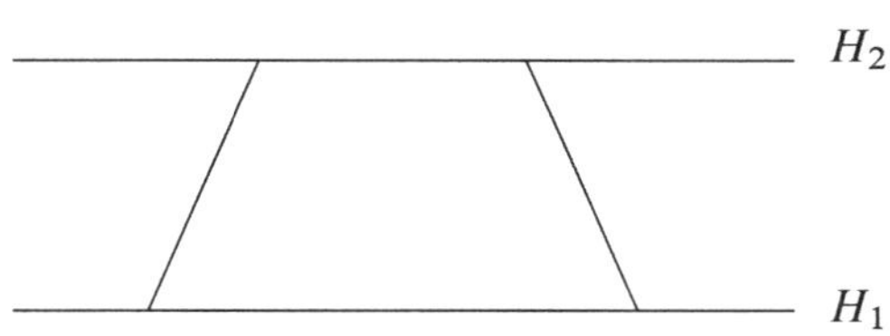

Figure 13

The following physical requirement is introduced.

Postulate. The energy-momentum tensor (gravitational stress) satisfies the following positivity condition. Let $T(\, ,\,)$ be the corresponding quadratic form in the tangent space at each point. Then we have

$$T(X, Y) \geq 0 \tag{237}$$

whenever $X,\ Y$ are future-directed timelike vectors at a point.

Strong version. We have equality only if the field is trivial at that point.

Remark. The postulate implies that $T(X,Y) \geq 0$ whenever $X,\ Y$ are future-directed non-spacelike vectors at a point.

If now $\mathcal{H}$ is a Cauchy hypersurface with unit future-directed timelike normal N, then

$$\int_{\mathcal{H}} {}^*P = \int_{\mathcal{H}} P^N \, d\mu_{\bar{g}} = \int_{\mathcal{H}} T(X,N)\, d\mu_{\bar{g}}, \tag{238}$$

where P^N is the N-component of P. That is, complementing $N = E_0$ with a frame (E_1, E_2, E_3) for $\mathcal{H}$ at a point to a frame for M at that point, we expand:

$$P = P^N\ N + \sum_{i=1}^{3} P^i\ E_i. \tag{239}$$

Then, since $g(N,N) = -1$, we have

$$P^N = -g\ (N,P) = T\ (X,N). \tag{240}$$

The postulate then implies that the integral (238) is ≥ 0 whenever X is non-spacelike future-directed.

We shall now give an example of a case where the above theory, which is based on the gravitational stress, does not apply, nevertheless Noether's theorem still applies.

Example (Electromagnetic theory in a medium). Consider a medium at rest in some Lorentz frame, the properties of which are invariant under the corresponding time translations. That is, there is a parallel timelike future-directed unit vector field u which is the material 4-velocity, and the material properties are invariant under the group generated by u. Then there are rectangular coordinates (x^0, x^1, x^2, x^3) such that

$$u = \frac{\partial}{\partial x^0} \quad (x^0 = t)$$

and

$$L = L\ (x^i, E^i, B^i;\ i = 1,2,3),$$

where

$$E^i = F^{0i}, \quad F_{ij} = \epsilon_{ijk}\ B^k$$

$({}^*F^{0i} = -B^i)$. As usual, $F = dA$, or, in components, $F_{\mu\nu} = \partial_\mu A_\nu - \partial_\nu A_\mu$. The E^i and B^i are the components of the electric and magnetic field respectively. The corresponding displacements are defined by

$$D^i = -\frac{\partial L}{\partial E^i},$$
$$H^i = \frac{\partial L}{\partial B^i}.$$

The Euler–Lagrange equations are *Maxwell's equations* (in the absence of charges and currents):

$$\nabla \cdot D = 0,$$
$$\nabla \times H - \frac{\partial D}{\partial t} = 0.$$

We also have the condition $dF = 0$ which reads

$$\nabla \cdot B = 0,$$
$$\nabla \times E + \frac{\partial B}{\partial t} = 0.$$

The simplest case is that of a homogeneous and isotropic medium.

Homogeneous medium. L is invariant under space translations:

$$\frac{\partial L}{\partial x^i} = 0, \tag{241}$$

thus

$$L = L(E^i, B^i : i = 1, 2, 3). \tag{242}$$

Isotropic medium. L is invariant under space rotations.

If the medium is *homogeneous and isotropic*, then we have

$$L = L\,(\,|E|^2,\ |B|^2,\ E \cdot B\,). \tag{243}$$

Noting that $\alpha = F^{\mu\nu} F_{\mu\nu} = 2(-|E|^2 + |B|^2)$ and $\beta = F^{\mu\nu}\, {}^*F_{\mu\nu} = 4E \cdot B$, we see that even in this case the Lagrangian is more general than that of Example 2. This is due to the fact that the vector field u is an additional structure on M, besides the metric g.

Nevertheless, Noether's theorem still applies to the general medium above. The invariance under time translations gives rise to the conservation of energy:

$$\mathcal{E} = \int_{\Sigma_t} \varepsilon\, d^3x, \tag{244}$$

where $\varepsilon = E \cdot D + L$ is the energy density and Σ_t is the hyperplane $x^0 = t$. In fact, we have the differential conservation law

$$\frac{\partial \varepsilon}{\partial t} + \nabla \cdot f = 0, \tag{245}$$

where f is the energy flux,

$$f = E \times H, \tag{246}$$

the vector field

$$\varepsilon \frac{\partial}{\partial x^0} + f \cdot \nabla \tag{247}$$

being the Noether current corresponding to time translations.

We shall confine ourselves in the following to geometric Lagrangians. Such a Lagrangian possesses the symmetries of the metric, therefore the pullback by an isometry of a solution of the Euler–Lagrange equations corresponding to a given metric is also a solution of the same equations. In the case that the Euler–Lagrange equations are linear, the difference of two solutions is also a solution. It follows that the Lie derivative of a solution with respect to a vector field generating a 1-parameter group of isometries is also, in the linear case, a solution, being the limit of a difference quotient. Moreover, if the action is conformally invariant, the same applies to the case of a vector field generating a 1-parameter group of conformal isometries.

Given a field Ψ, consider the derived fields

$$\Psi_n = \mathcal{L}_{Y_{i_1}} \dots \mathcal{L}_{Y_{i_n}} \Psi \tag{248}$$

with $i_1, \dots, i_n = \{1, \dots, m\}$, and $\{Y_1, \dots, Y_m\}$ a set of generators of the m parameter subgroup of the isometry (or conformal isometry) group of g (Killing or conformal Killing vector fields of g). The construction of the previous paragraph gives a positive conserved quantity associated to Ψ_n and to each vector field X generating a 1-parameter subgroup of the isometry (or conformal isometry) group of g which is, moreover, non-spacelike and future-directed. If we have enough positive conserved quantities of this type, then Sobolev inequalities imply the uniform decay of solutions.

In the non-linear case the Lie derivative with respect to a Killing vector field (or conformal Killing vector field) is no longer a solution of the original Euler–Lagrange equations but it is a solution of the equations of variation, namely the Euler–Lagrange equations corresponding to the linearized Lagrangian about the given solution. The underlying structure on which this linearized Lagrangian depends is not only the metric but also the background solution. Thus the construction which we have discussed will not yield a conserved quantity unless the background solution is invariant under the 1-parameter group in question. In general, we will have error terms which involve the Lie derivative of the background solution with respect to the corresponding vector field. The difference of the quantity corresponding to a spacelike or null hypersurface from the same quantity corresponding to the initial hypersurface will be the integral of the error terms over the spacetime domain bounded by these two hypersurfaces.

The aim in the non-linear case is then to achieve closure: obtain enough positive quantities such that the error integrals can be bounded in terms of the quantities themselves. Once closure is achieved the global existence theorem for small initial data will follow by a continuity argument.

We now consider in more detail the conformal group of Minkowski spacetime.

Conformal group of Minkowski spacetime. This group consists of:

1. *Spacetime translations.* These form an Abelian group. They are generated by the vector fields (rectangular coordinates)
$$T_\mu = \frac{\partial}{\partial x^\mu}, \quad \mu = 0, 1, 2, 3, \tag{249}$$
of degree -1. T_μ generates translations along the μ-th coordinate axis.
2. *Spacetime rotations* (Lorentz transformations). These constitute the Lorentz group SO(3, 1). They are generated by the vector fields
$$\Omega_{\mu\nu} = x_\mu \frac{\partial}{\partial x^\nu} - x_\nu \frac{\partial}{\partial x^\mu}, \quad \mu, \nu = 0, 1, 2, 3, \;\; \mu < \nu, \tag{250}$$
of degree 0. Here
$$x_\mu = \eta_{\mu\alpha} x^\alpha.$$
$\Omega_{\mu\nu}$ generates rotations in the μ, ν-coordinate plane.
3. *Scale transformations* ($x \mapsto ax$, $a > 0$). They are generated by the vector field
$$S = x^\mu \frac{\partial}{\partial x^\mu} \tag{251}$$
of degree 0, which commutes with the spacetime rotations.
4. *Inverted spacetime translations.* These also form an Abelian group. They are generated by the vector fields
$$K_\mu = -2\, x_\mu\, S + (x, x)\, T_\mu, \quad \mu = 0, 1, 2, 3, \tag{252}$$
of degree 1. Here
$$(x, x) = \eta_{\alpha\beta}\, x^\alpha\, x^\beta.$$
The K_μ are the generators of the 1-parameter groups
$$I P_\mu I, \tag{253}$$

where P_μ is a translation along the μ-coordinate axis

$$P_\mu\, x = (x^0, \ldots, x^\mu + t, \ldots, x^3),$$

and I is the *inversion*, a discrete conformal isometry of Minkowski spacetime, defined below.

The *inversion* I is defined by

$$I : \tilde{x} \mapsto x = \frac{\tilde{x}}{(\tilde{x}, \tilde{x})}, \quad ((x, y) = \eta_{\mu\nu} x^\mu y^\nu). \tag{254}$$

Note that

$$(\tilde{x}, \tilde{x})(x, x) = 1.$$

The inverse is

$$I^{-1} : x \mapsto \tilde{x} = \frac{x}{(x, x)}. \tag{255}$$

Thus

$$I^{-1} = I \quad \text{or} \quad I \circ I = \mathrm{id}.$$

We have

$$\frac{\partial x^\mu}{\partial \tilde{x}^\alpha} = \frac{\delta^\mu_\alpha}{(\tilde{x}, \tilde{x})} - \frac{2\tilde{x}^\mu \tilde{x}_\alpha}{(\tilde{x}, \tilde{x})^2}.$$

The pullback by I of the Minkowski metric η then reads

$$\begin{aligned} (I^* \eta_{\alpha\beta})\, (\tilde{x}) &= \frac{\partial x^\mu}{\partial \tilde{x}^\alpha} \frac{\partial x^\nu}{\partial \tilde{x}^\beta}\, \eta_{\mu\nu}\, (x) \\ &= \Omega^{-2}\, (\tilde{x}) \cdot \eta_{\alpha\beta}, \end{aligned}$$

where

$$\Omega(\tilde{x}) = -(\tilde{x}, \tilde{x}).$$

Thus we have

$$I^*\, \eta = \Omega^{-2}\, \eta, \tag{256}$$

that is, I is a conformal isometry of η.

We shall restrict the inversion mapping to $I^+(0)$ in the x-coordinates. ($I^+(0)$ is the chronological future of the origin.) Then (x, x), $(\tilde{x}, \tilde{x}) < 0$. So, $\Omega > 0$.

Proposition 7. *The inversion maps $I^+(0)$ (in x) to $I^-(0)$ (in $\tilde{x}$). In fact, the inversion maps light cones onto light cones.*

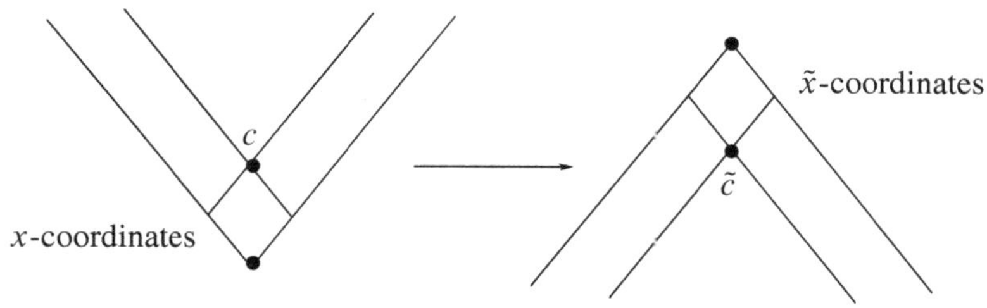

Figure 14

Proof. The light cone with vertex c is given by

$$(x-c,\ x-c)=0.$$

Then we have

$$\begin{aligned}(\tilde{x}-\tilde{c},\ \tilde{x}-\tilde{c}) &= (\tilde{x},\ \tilde{x}) - 2\,(\tilde{x},\ \tilde{c}) + (\tilde{c},\ \tilde{c}) \\ &= \frac{1}{(x,x)} - 2\,\frac{(x,c)}{(x,x)(c,c)} + \frac{1}{(c,c)} \\ &= \frac{(x-c,x-c)}{(x,x)(c,c)} = 0.\end{aligned}$$

This proves the proposition. □

In particular, future light cones are mapped onto future light cones and past light cones onto past light cones.

Remark. The entire causal future of a point $c \in I^+(0)$ is mapped onto a bounded region in $I^-(0)$.

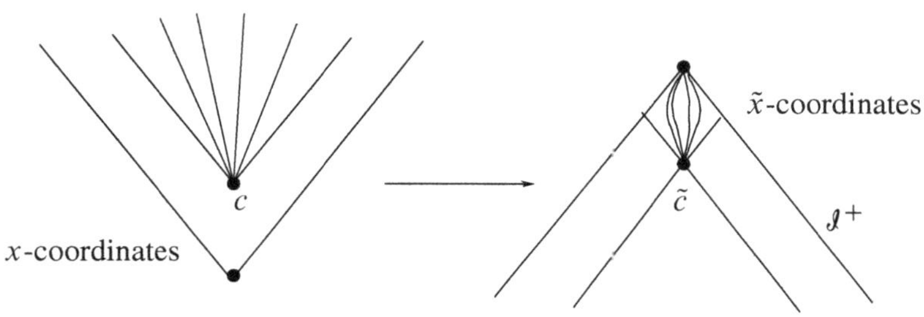

Figure 15

The infinity in $I^+(0)$ (in x) is mapped onto $\partial I^-(0)$ (in $\tilde{x}$). The cone $\partial I^-(0)\backslash 0$ is the *future null infinity* denoted by $\mathcal{J}^+$, a concept introduced by R. Penrose (see [24]), and the origin 0 is the future timelike infinity. Any null geodesic in x is mapped onto a null geodesic in $\tilde{x}$ with a future end-point on $\mathcal{J}^+$. Any timelike geodesic in x is mapped onto a timelike curve in $\tilde{x}$ ending at 0.

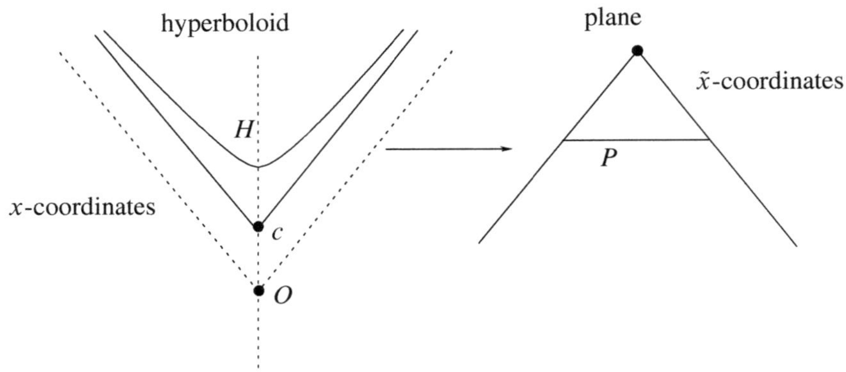

Figure 16

The inverse image of a spacelike hyperplane P,

$$\tilde{t} = -k, \quad k \text{ a positive constant}$$

$$\Big(\tilde{t} = \tilde{x}^0;\ \tilde{r} = \sqrt{\textstyle\sum_{i=1}^3 (\tilde{x}_i)^2}\ \Big),$$

is a spacelike hyperboloid H:

$$t = \frac{1}{2k} + \sqrt{\left(\frac{1}{2k}\right)^2 + r^2}$$

$$\Big(t = x^0,\ r = \sqrt{\textstyle\sum_{i=1}^3 (x^i)^2}\ \Big)$$

This is a hyperboloid through $x^0 = \frac{1}{k}$ on the x^0-axis, which is asymptotic to $\partial I^+(c)$, where c is the point $x^0 = \frac{1}{2k}$ on the x^0 axis. This hyperboloid is intrinsically a space of constant negative curvature $-(2k)^2$.

We shall consider examples where the initial data have compact support as well as an example where the initial data have non-compact support.

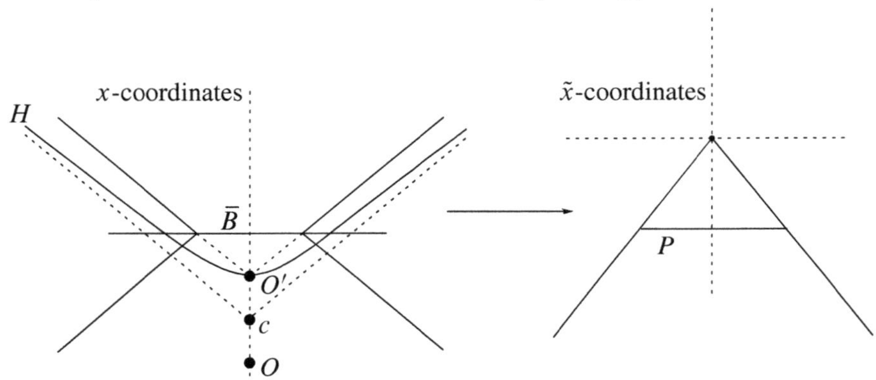

Figure 17

Given initial data of compact support, let $\overline{B}$ be the smallest closed ball on the initial hyperplane containing the support of the data. We choose the x^0-axis to be the straight line in Minkowski space orthogonal to the initial hyperplane through the center of $\overline{B}$.

Let O' be the point on the x^0-axis whose future light cone intersects the initial hyperplane at $\partial\overline{B}$. We then translate the origin of the (rectangular) spacetime coordinates to a point O on the x^0-axis which lies properly to the past of the point O' and we perform the inversion mapping relative to this new origin. We define the positive constant k so that $\frac{1}{k}$ is equal to the interval OO' and consider the hyperplane $P : \tilde{t} = -k$ in the $\tilde{x}$-space and the corresponding hyperboloid H in the x-space. H passes through the point O'.

We suppose that our system of equations is derived from a Lagrangian and that it admits a trivial global solution, at which it is regularly hyperbolic, with characteristic cones coinciding with the light cones of the underlying Minkowski spacetime.

Then, if the initial data is sufficiently close to the trivial data, a solution will exist in the closed spacetime slab (in the original x-space) bounded in the future by the initial hyperplane and in the past by the parallel hyperplane through O'. Moreover, by the domain of dependence theorem, the solution is trivial outside the union of the causal future and causal past of $\overline{B}$, in particular in the non-compact portion of H lying to the future of the initial hyperplane.

We shall investigate below the conditions under which the Lagrangian transforms under the inversion map into a regular Lagrangian in the $\tilde{x}$-space. The corresponding system of Euler–Lagrange equations will then be a regular system in the $\tilde{x}$-space equivalent to the original system.

We can then consider the Cauchy problem in the $\tilde{x}$-space with initial data on the hyperplane P coming through the inversion mapping from the induced data on H. This initial data is trivial in a neighborhood of the intersection $P \cap \mathcal{I}^+$, therefore extends trivially on P outside this intersection. If the initial data on P is sufficiently close to trivial, which is the case if we require the original data (in x-space) to be suitably close to trivial, then the solution of the transformed system in $\tilde{x}$-space will exist in the entire closed slab bounded in the past by P and in the future by the parallel hyperplane through the origin. Transforming then the solution back to the original x-space, we obtain a global solution for the given initial data. In this way a global existence theorem for small initial data is proven. This approach was first generally expounded in [10].

Setting

$$g = I^* \eta,$$

the metric $\tilde{g}$, given by

$$\tilde{g} = \Omega^2 g, \tag{257}$$

coincides, according to (256), with the Minkowski metric η:

$$\tilde{g} = \eta.$$

Note that $\Omega = 0$ on $\partial I^-(0) = \mathcal{I}^+ \cup 0$ in $\tilde{x}$-space, which corresponds to infinity in the x-space. In the case of a scalar field Φ, we also set

$$\widetilde{\Phi} = \Omega^{-1}\,\Phi. \tag{258}$$

Then we have

$$\begin{aligned}\sigma &= g^{\mu\nu}\,\partial_\mu\,\Phi\,\partial_\nu\,\Phi\\ &= \Omega^2\,\tilde{g}^{\mu\nu}\,\partial_\mu\,(\Omega\,\widetilde{\Phi})\,\partial_\nu\,(\Omega\,\widetilde{\Phi})\\ &= \Omega^4\,\tilde{g}^{\mu\nu}\,(\partial_\mu\,\widetilde{\Phi} + \widetilde{\Phi}\,\partial_\mu\log\Omega)\,(\partial_\nu\,\widetilde{\Phi} + \widetilde{\Phi}\,\partial_\nu\log\Omega).\end{aligned}$$

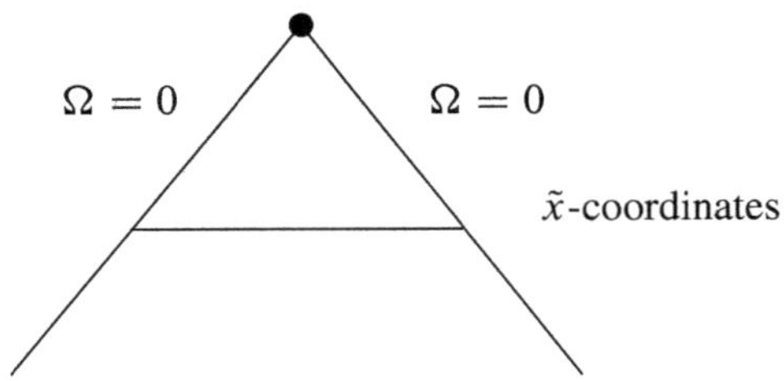

Figure 18

Consider the Lagrangian

$$L = \sigma\,d\mu_g.$$

The corresponding Euler–Lagrange equation is the linear wave equation for Φ in the metric g:

$$\Box\,\Phi = 0.$$

Since

$$d\mu_g = \Omega^{-4}\,d\mu_{\tilde{g}},$$

we then have

$$\begin{aligned}L &= \tilde{g}^{\mu\nu}\,(\partial_\mu\,\widetilde{\Phi} + \widetilde{\Phi}\,\partial_\mu\log\Omega)\,(\partial_\nu\,\widetilde{\Phi} + \widetilde{\Phi}\,\partial_\nu\log\Omega)\,d\mu_{\tilde{g}}\\ &= (\tilde{\sigma} + 2\,\tilde{g}^{\mu\nu}\,\widetilde{\Phi}\,\partial_\mu\,\widetilde{\Phi}\,\partial_\nu\log\Omega + \tilde{g}^{\mu\nu}\,\widetilde{\Phi}^2\,\partial_\mu\log\Omega\,\partial_\nu\log\Omega)\,d\mu_{\tilde{g}}\end{aligned}$$

where

$$\tilde{\sigma} = \tilde{g}^{\mu\nu}\,\partial_\mu\,\widetilde{\Phi}\,\partial_\nu\,\widetilde{\Phi}.$$

Now

$$\begin{aligned}2\,\tilde{g}^{\mu\nu}\,\widetilde{\Phi}\,\partial_\mu\,\widetilde{\Phi}\,\partial_\nu\log\Omega &= \partial_\mu\,(\widetilde{\Phi}^2)\,\tilde{g}^{\mu\nu}\,\partial_\nu\log\Omega\\ &= \tilde{\nabla}_\mu\,(\widetilde{\Phi}^2\,\tilde{g}^{\mu\nu}\,\partial_\nu\log\Omega) - \widetilde{\Phi}^2\,\tilde{\Box}\log\Omega\,.\end{aligned}$$

Consider the vector field

$$V^\mu = \tilde{\Phi}^2\, \tilde{g}^{\mu\nu} \partial_\nu \log \Omega. \tag{259}$$

Then

$$\tilde{\nabla}_\mu\, V^\mu\, d\mu_{\tilde{g}} \tag{260}$$

is a *null Lagrangian*: we can subtract it from L without affecting the Euler–Lagrange equations. After this subtraction the Lagrangian becomes

$$\tilde{L} = \{\tilde{\sigma} + \tilde{\Phi}^2\, (-\tilde{\Box} \log \Omega + \tilde{g}^{\mu\nu}\, \partial_\mu \log \Omega\, \partial_\nu \log \Omega)\}\, d\mu_{\tilde{g}}. \tag{261}$$

Here

$$-\tilde{\Box} \log \Omega + \tilde{g}^{\mu\nu}\, \partial_\mu \log \Omega\, \partial_\nu \log \Omega = \Omega\, \tilde{\Box}\, \Omega^{-1} = 0. \tag{262}$$

For,

$$\tilde{\Box}\, \Omega^{-1} = \eta^{\mu\nu}\, \frac{\partial^2}{\partial \tilde{x}^\mu \partial \tilde{x}^\nu} \Big(- \frac{1}{(\tilde{x}, \tilde{x})} \Big) = 0.$$

Therefore, we obtain

$$\tilde{L} = \tilde{\sigma}\, d\mu_{\tilde{g}}.$$

Thus, the Euler–Lagrange equation is equivalent to the linear wave equation for $\tilde{\Phi}$ in the metric $\tilde{g}$:

$$\tilde{\Box}\, \tilde{\Phi} = 0.$$

Example 1 (Non-linear wave equation in Minkowski spacetime). Consider now a general Lagrangian of the form

$$L^* = L^*\, (\sigma) \quad (L = L^*\, d\mu_g).$$

The corresponding Euler–Lagrange equation is the non-linear wave equation

$$\nabla_\mu (G(\sigma) \partial^\mu \Phi) = 0 \quad \text{where } G = \frac{dL^*}{d\sigma}. \tag{263}$$

By subtraction of an appropriate constant from L^* and multiplication by another such constant we arrive at L^* of the form

$$L^*(\sigma) = \sigma + \sigma^2 F(\sigma),$$

where F is a smooth function in a neighbourhood of 0. The contribution to L of the non-linear term in L^* is

$$\begin{aligned}\sigma^2 F(\sigma)\, d\mu_g &= (\tilde{g}^{\mu\nu}\, \partial_\mu\, (\Omega\, \tilde{\Phi})\, \partial_\nu\, (\Omega\, \tilde{\Phi}))^2 F(\Omega^2\, \tilde{g}^{\mu\nu}\, \partial_\mu\, (\Omega\, \tilde{\Phi})\, \partial_\nu\, (\Omega\, \tilde{\Phi}))\, d\mu_{\tilde{g}} \\ &=: N\, d\mu_{\tilde{g}}.\end{aligned}$$

The total Lagrangian is thus equivalent to

$$(\tilde{\sigma} + N)\, d\mu_{\tilde{g}},$$

where N is regular at $\Omega = 0$. In fact, we have

$$N|_{\Omega=0} = (\tilde{g}^{\mu\nu}\, \partial_\mu\, \Omega\, \partial_\nu\, \Omega)^2 \cdot \tilde{\Phi}^2\, F\,(0) = 0$$

because the hypersurface $\Omega = 0$, i.e. $\mathcal{I}^+$ is null:

$$\tilde{g}^{\mu\nu} \partial_\mu \Omega \partial_\nu \Omega = 0 \quad \text{along the hypersurface } \Omega = 0.$$

A particular example of the above kind is the following.

Example (Minimal timelike surfaces in 5-dimensional Minkowski spacetime). Consider the Minkowski metric on $\mathbb{R}^5$:

$$-(dx^0)^2 + \sum_{i=1}^{3} (dx^i)^2 + (dx^4)^2. \tag{264}$$

Then an arbitrary graph of x^4 over the 4-dimensional Minkowski spacetime is of the form

$$x^4 = \Phi\,(x^0,\, x^1,\, x^2,\, x^3).$$

The area element of such a graph is

$$\sqrt{1 + \sigma}\, d^4x,$$

and the equation of minimal surfaces is the Euler–Lagrange equation corresponding to this Lagrangian.

Example 2 (Non-linear electrodynamics in Minkowski spacetime). Let

$$\alpha = F^{\mu\nu}\, F_{\mu\nu}, \quad \beta = F^{\mu\nu}\, F^*_{\mu\nu}$$

and

$$L = L^*\,(\alpha,\, \beta)\, d\mu_g. \tag{265}$$

We stipulate that under a conformal transformation $g \mapsto \tilde{g} = \Omega^2 g$, the 2-form

$$F = \sum_{\mu<\nu} F_{\mu\nu}\, dx^\mu \wedge dx^\nu \tag{266}$$

remains unchanged. Thus, we have

$$\begin{aligned} \alpha &= g^{\mu\kappa}\, g^{\nu\lambda}\, F_{\kappa\lambda}\, F_{\mu\nu} = \Omega^4\, \underbrace{\tilde{g}^{\mu\kappa}\, \tilde{g}^{\nu\lambda}\, F_{\kappa\lambda}\, F_{\mu\nu}}_{=\tilde{\alpha}} \\ &= \Omega^4\, \tilde{\alpha}. \end{aligned}$$

Expressing

$$F^*_{\mu\nu} = \frac{1}{2}\, F^{\kappa\lambda}\, \epsilon_{\kappa\lambda\mu\nu} = \frac{1}{2}\, g^{\kappa\rho}\, g^{\lambda\sigma}\, F_{\rho\sigma}\, \epsilon_{\kappa\lambda\mu\nu},$$

and taking into account the fact that $\epsilon_{\kappa\lambda\mu\nu}$, the volume form of g, is related to $\tilde{\epsilon}_{\kappa\lambda\mu\nu}$, the volume form of $\tilde{g}$, by

$$\epsilon_{\kappa\lambda\mu\nu} = \Omega^{-4}\, \tilde{\epsilon}_{\kappa\lambda\mu\nu},$$

we see that

$$F^*_{\mu\nu} = \frac{1}{2}\, \tilde{g}^{\kappa\rho}\, \tilde{g}^{\lambda\sigma}\, F_{\rho\sigma}\, \tilde{\epsilon}_{\kappa\lambda\mu\nu} = F^{\tilde{*}}_{\mu\nu}.$$

Therefore,

$$\begin{aligned} \beta &= g^{\mu\kappa}\, g^{\nu\lambda}\, F_{\kappa\lambda}\, F^*_{\mu\nu} \\ &= \Omega^4\, \tilde{g}^{\mu\kappa}\, \tilde{g}^{\nu\lambda}\, F_{\kappa\lambda}\, F^{\tilde{*}}_{\mu\nu} \\ &= \Omega^4\, \tilde{\beta}. \end{aligned}$$

Subtracting from L^* an appropriate constant we can bring L^* to the form

$$L^* = \alpha\, G\,(\alpha, \beta) + \beta\, H\,(\alpha, \beta),$$

where G and H are smooth functions in a neighbourhood of $(0, 0)$. Thus,

$$L = \{\alpha\, G\,(\alpha, \beta) + \beta\, H\,(\alpha, \beta)\} d\mu_g.$$

Since $d\mu_g = \Omega^{-4} d\mu_{\tilde{g}}$, we then have

$$L = \{\tilde{\alpha}\, G\,(\Omega^4\tilde{\alpha}, \Omega^4\tilde{\beta}) + \tilde{\beta}\, H\,(\Omega^4\tilde{\alpha}, \Omega^4\tilde{\beta})\}\, d\mu_{\tilde{g}}.$$

This is regular at $\Omega = 0$. In fact,

$$L|_{\Omega=0} = \{\tilde{\alpha}\, G\,(0, 0) + \tilde{\beta}\, H\,(0, 0)\}\, d\mu_{\tilde{g}}.$$

For a general electromagnetic Lagrangian, the *displacement* $G^{\mu\nu}$ is defined by

$$G^{\mu\nu} = \frac{\partial L^*}{\partial F_{\mu\nu}}. \tag{267}$$

(Note: In taking the partial derivative with respect to $F_{\mu\nu}$, the equality $F_{\nu\mu} = -F_{\mu\nu}$ is to be taken into account. Thus, $\lim_{t\to 0} t^{-1}\{L^*(F + t\dot{F}) - L^*(F)\} = \frac{1}{2}\frac{\partial L^*}{\partial F_{\mu\nu}}\dot{F}_{\mu\nu}$.) So, for the Lagrangian giving rise to the linear Maxwell equations,

$$\begin{aligned} L^* &= \frac{1}{4}\alpha, \\ G^{\mu\nu} &= F^{\mu\nu}. \end{aligned}$$

In general, in terms of a splitting into time and space (as given by the choice of a time coordinate x^0):

$$\begin{aligned} F^{0i} &= E^i && \text{is the electric field,} \\ F_{ij} &= \epsilon_{ijk} B^k && \text{is the magnetic field} \end{aligned}$$

($F^{*0i} = -B^i$, $F^*_{ij} = \epsilon_{ijk} E^k$), and we have

$$\alpha = -2|E|^2 + 2|B|^2, \quad \beta = 2E \cdot B.$$

Then with

$$D^i = -\frac{\partial L^*}{\partial E_i}, \tag{268}$$

$$H^i = \frac{\partial L^*}{\partial B_i}, \tag{269}$$

the displacement reads:

$$\begin{aligned} G^{0i} &= D^i && \text{electric displacement,} \\ G_{ij} &= \epsilon_{ijk} H^k && \text{magnetic displacement} \end{aligned}$$

($G^{*0i} = -H^i$, $G^*_{ij} = \epsilon_{ijk} D^k$). Also, setting

$$\gamma = G^{\mu\nu} G_{\mu\nu}, \quad \delta = G^{\mu\nu} G_{*\mu\nu} \tag{270}$$

we have

$$\gamma = -2|D|^2 + 2|H|^2, \quad \delta = 2D \cdot H.$$

Let us consider the mapping

$$F \mapsto G \tag{271}$$

We have

$$\frac{\partial \alpha}{\partial F_{\mu\nu}} = 4\, F^{\mu\nu}, \quad \frac{\partial \beta}{\partial F_{\mu\nu}} = 4\, F^{*\mu\nu},$$

hence

$$G^{\mu\nu} = 4\, \frac{\partial L^*}{\partial \alpha}\, F^{\mu\nu} + 4\, \frac{\partial L^*}{\partial \beta}\, F^{*\mu\nu}. \tag{272}$$

(Note that $F^{**}_{\mu\nu} = -F_{\mu\nu}$.) It follows that the mapping

$$(\alpha,\ \beta) \mapsto (\gamma,\ \delta) \tag{273}$$

is given by

$$\begin{aligned}\gamma &= \left[\left(\frac{\partial L^*}{\partial \alpha}\right)^2 - \left(\frac{\partial L^*}{\partial \beta}\right)^2\right]\alpha + 2\,\frac{\partial L^*}{\partial \alpha}\,\frac{\partial L^*}{\partial \beta}\,\beta,\\ \delta &= -2\,\frac{\partial L^*}{\partial \alpha}\,\frac{\partial L^*}{\partial \beta}\,\alpha + \left[\left(\frac{\partial L^*}{\partial \alpha}\right)^2 - \left(\frac{\partial L^*}{\partial \beta}\right)^2\right]\beta.\end{aligned} \tag{274}$$

The Maxwell equations

$$d\,F = 0 \quad (F = d\,A) \tag{275}$$

take the form

$$\nabla \cdot B = 0, \tag{276}$$

$$\nabla \times E + \frac{\partial B}{\partial t} = 0. \tag{277}$$

The remaining Maxwell equations are the Euler–Lagrange equations

$$d\,G^* = 0. \tag{278}$$

These take the form

$$\nabla \cdot D = 0, \tag{279}$$

$$\nabla \times H - \frac{\partial D}{\partial t} = 0. \tag{280}$$

The initial conditions consist of the specification of B and D at $t = 0$ subject to the constraint equations

$$\nabla \cdot B = 0, \tag{281}$$

$$\nabla \cdot D = 0. \tag{282}$$

Initial data (B_0, D_0) which are C^∞ and of compact support can readily be constructed by taking curls of C^∞ vector fields of compact support.

Example 3 (Gauge theory of a complex line bundle over Minkowski spacetime). We have a complex line bundle over Minkowski spacetime, equipped with a Hermitian metric. The bundle is topologically trivial, being diffeomorphic to the product $\mathbb{R}^4 \times \mathbb{C}$. Let us choose an orthonormal basis section σ, that is $|\sigma| = 1$. Here $|\cdot|$ is the norm corresponding to the Hermitian inner product on each complex line fibre. As each fibre has only one complex dimension, only the condition $|\sigma| = 1$ has to be fulfilled. The imaginary line being the Lie algebra of $U(1)$, one has connection coefficients iA_μ defined by the covariant derivative

$$D_\mu \sigma = i A_\mu \sigma.$$

The wave function is a section s of the complex line bundle,

$$s = \Phi\sigma,$$

where Φ is a complex-valued function in Minkowski spacetime. We have

$$|s| = |\Phi|,$$

where $|\Phi(x)|$ is the absolute value of the complex number $\Phi(x)$. We then find

$$D_\mu\, s = (D_\mu\, \Phi)\, \sigma,$$

where

$$D_\mu\, \Phi = \partial_\mu\, \Phi + i\, A_\mu\, \Phi.$$

The Lagrangian of the theory reads

$$L = L^*\, d\mu_g$$

with

$$L^* = \frac{1}{2}\, D_\mu\, \Phi\, \overline{D^\mu \Phi} + \frac{\lambda}{4}\, |\Phi|^4 + \frac{1}{4}\, F^{\mu\nu}\, F_{\mu\nu}. \tag{283}$$

Here, $F = dA$, that is $F_{\mu\nu} = \partial_\mu A_\nu - \partial_\nu A_\mu$, iF being the curvature of the bundle. The 1-form A is identified with the electromagnetic potential and the 2-form F with the electromagnetic field. The theory describes a charged scalar field in interaction with the electromagnetic field. The Euler–Lagrange equations are

$$\begin{aligned} \nabla_\nu\, F^{\mu\nu} &= J^\mu = \operatorname{Im}(\bar{\Phi}\, D^\mu\, \Phi), \\ D^\mu\, D_\mu\, \Phi &= \lambda\, |\Phi|^2\, \Phi. \end{aligned} \tag{284}$$

Here, J^μ is the electric current density. Under change of basis section

$$\sigma \mapsto e^{i\chi}\, \sigma,$$

we have

$$\begin{aligned} \Phi &\mapsto e^{-i\chi}\, \Phi, \\ A_\mu &\mapsto A_\mu + \partial_\mu\, \chi. \end{aligned} \tag{285}$$

The last pair of transformations are the gauge transformations.

Proposition 8. *The equivalence class of L is conformally invariant.*

Proof. The last term in (283) is

$$\frac{1}{4}\,\alpha\, d\mu_g$$

which we have shown to be conformally invariant, being equal to

$$\frac{1}{4}\,\tilde{\alpha}\, d\mu_{\tilde{g}}.$$

The middle term in (283) is also conformally invariant, being equal to

$$\frac{1}{4}\,|\tilde{\Phi}|^4\, d\mu_{\tilde{g}}; \quad \tilde{\Phi} = \Omega^{-1}\Phi.$$

The first term is $\frac{1}{2}\sigma d\mu_g$, where now

$$\begin{aligned}
\sigma &= g^{\mu\nu}\, D_\mu\, \Phi\, \overline{D_\nu\, \Phi} \\
&= \Omega^2\, \tilde{g}^{\mu\nu}\, D_\mu(\Omega\tilde{\Phi})\, \overline{D_\nu(\Omega\tilde{\Phi})} \\
&= \Omega^4\ \tilde{g}^{\mu\nu}\, (D_\mu\, \tilde{\Phi} + \tilde{\Phi}\, \partial_\mu \log\Omega)\, (\overline{D_\nu\tilde{\Phi}} + \bar{\tilde{\Phi}}\, \partial_\nu \log\Omega) \\
&= \Omega^4\, \tilde{\sigma} + \tilde{\nabla}_\mu\, V^\mu, \\
V^\mu &= \tilde{g}^{\mu\nu}|\tilde{\Phi}|^2\partial_\nu \log\Omega
\end{aligned}$$

(see (262)). Therefore,

$$\frac{1}{2}\,\sigma\, d\mu_g \quad \text{is equivalent to} \quad \frac{1}{2}\,\tilde{\sigma}\, d\mu_{\tilde{g}}$$

and the proposition is proven. □

The initial conditions consist of the specification of Φ, $D_0\Phi$, E, B at $t = 0$. These are subject to the constraint equations

$$\begin{aligned}
\nabla\cdot B &= 0, \\
\nabla\cdot E &= \rho,
\end{aligned} \tag{286}$$

where

$$\rho = J^0 = \operatorname{Im}(\bar{\Phi}\, D^0\, \Phi)$$

is the electric charge density. Here we can take all the data *except* E to have compact support. Writing

$$E = U + \nabla\, h,$$

where U is a C^∞ compactly supported divergence-free vector field, $\nabla\cdot U = 0$, the function h must satisfy $\Delta h = \rho$. Therefore h is a harmonic function outside the ball of support of the rest of the initial data. Outside the future of this ball in Minkowski spacetime we have

$$
\begin{aligned}
\Phi &= 0, \\
B &= 0, \\
E &= \nabla h, \quad \text{independent of } t.
\end{aligned}
\tag{287}
$$

For, (287) is a solution of (284). By the domain of dependence theorem it must be *the* solution in the domain in question corresponding to the given initial data.

In 3-dimensional Euclidean space $(\mathbb{R}^3, e)$ we consider the inversion map

$$x \mapsto x' = i^{-1}(x) = \frac{x}{r^2},$$

where $r = |x|$. With $r' = |x'|$, we have $r' = r^{-1}$. Then

$$i^* e = \Omega^{-2}\, e,$$

where $\Omega = r'^2$. Thus, setting $g = i^* e$, $g' = e$, we have $g' = \Omega^2 g$. Moreover, setting also $h' = \Omega^{-\frac{1}{2}} h$, we have

$$\triangle\, h = 0 \iff \triangle'\, h' = 0.$$

Now our function h is harmonic in $\mathbb{R}^3 \backslash \overline{B}_R$, where $\overline{B}_R$ is the smallest closed ball containing the support of the data for Φ, $D_0\, \Phi$, B, U and thus the support of ρ. It follows that h' is analytic in $B_{R'}$, where $R' = R^{-1}$. We have a convergent Taylor expansion at the origin, which represents the infinity of the original Euclidean space. This Taylor expansion

$$h' = a + b_i\, x'^i + c_{ij}\, x'^i\, x'^j + \cdots$$

(with $\Delta' h' = 0 \Leftrightarrow \operatorname{tr} c' = 0, \dots$) corresponds to the *multipole expansion* of h.

Translating the origin in Minkowski spacetime to a point along the straight line orthogonal to the initial hyperplane through the center of the ball $\overline{B}_R$, a point lying properly to the past of the point the future light cone of which intersects the initial hyperplane at $\partial \overline{B}_R$, we see that the *spacetime* inversion map I^{-1} in the exterior of the causal future and past of $\overline{B}_R$, takes the static solution discussed above to a solution of the same equations in the image of this domain which admits an analytic extension through $\mathcal{I}^+$. The arguments outlined previously then apply yielding a global existence and decay theorem for small initial data. The decay of the original fields follows trivially from the boundedness of the transformed fields.

4.2 Sketch of the proof of the global stability of Minkowski spacetime

4.2.1 The problem in General Relativity – two main difficulties

1. The definition of the energy-momentum tensor appropriate to a geometric Lagrangian, namely by considering the variation of the action with respect to the underlying metric, clearly fails, because this variation vanishes for the gravitational Lagrangian
$$L = -\frac{1}{4} R \, d\mu_g .$$
This vanishing is the statement of the Euler–Lagrange equations for gravitation, namely the Einstein (vacuum) equations.

   An alternative approach is to appeal to Noether's theorem after subtracting an appropriate divergence relative to a background metric, as we have done in defining the total energy (E), linear momentum (P^i), angular momentum (J^i) and center of mass integrals (C^i). Among these the energy has been shown to be positive. But, the energy (a quantity which scales like length) gives us control on the solutions only after the isoperimetric constant, a dimensionless quantity, is assumed under control. Therefore, the energy cannot be used, by itself, to prove regularity.

2. A general spacetime has no symmetries: the conformal isometry group of a general spacetime is trivial. Therefore we have no conformal Killing vector fields at our disposal, to use, in conjunction with energy-momentum tensors, to construct integral conserved quantities.

4.2.2 Resolution of the first difficulty. The idea of how to overcome the first difficulty is based on the following analogy with Maxwell's equations of electromagnetic theory.

Our aim is to derive estimates for the spacetime curvature which will give us the necessary control on regularity. The idea is to consider the Bianchi identities

$$\nabla_{[\alpha} R_{\beta\gamma]\delta\epsilon} = 0, \quad [\alpha\beta\gamma] \text{ a cyclic permutation,} \tag{288}$$

as differential equations for the curvature, and the Einstein equations

$$R_{\mu\nu} := g^{\alpha\beta} R_{\alpha\mu\beta\nu} = 0 \tag{289}$$

as algebraic conditions on the curvature. Breaking the connection between the metric and the curvature, we define a *Weyl field* $W_{\alpha\beta\gamma\delta}$ on a given 4-dimensional spacetime manifold $(M, g_{\mu\nu})$ to be a tensor field with the same algebraic properties as the Weyl (or conformal) curvature tensor. Namely:

$W_{\beta\alpha\gamma\delta} = W_{\alpha\beta\delta\gamma} = -W_{\alpha\beta\gamma\delta}$	antisymmetry in the first two as well as in the last two indices,
$W_{\alpha[\beta\gamma\delta]} = 0$	cyclic condition,

and the trace condition

$$g^{\alpha\beta}\, W_{\alpha\mu\beta\nu} = 0,$$

the analogue of the Einstein equations. The first two properties imply the symmetry under exchange of the first and second pair of indices: $W_{\gamma\delta\alpha\beta} = W_{\alpha\beta\gamma\delta}$.

Given a Weyl field W we can define a *right dual* W^* as well as a *left dual* ${}^*W$. The *left dual* is defined as

$$ {}^*W_{\alpha\beta\gamma\delta} = \frac{1}{2}\, \epsilon_{\mu\nu\alpha\beta}\, W^{\mu\nu}{}_{\gamma\delta} \tag{290}$$

by freezing the second pair of indices and considering W as a 2-form relative to the first pair. The *right dual* is defined as

$$ W^*_{\alpha\beta\gamma\delta} = \frac{1}{2}\, W_{\alpha\beta}{}^{\mu\nu}\, \epsilon_{\mu\nu\gamma\delta} \tag{291}$$

by freezing the first pair of indices and considering W as a 2-form in the second pair.

However, by virtue of the algebraic properties of W the two duals coincide:

$$ {}^*W = W^*. \tag{292}$$

We shall thus write only ${}^*W$ in the following. Moreover, ${}^*W$ is also a Weyl field. In fact, the cyclic condition for ${}^*W$ is equivalent (modulo the other conditions) to the trace condition for W and vice versa.

A Weyl field is subject, in the absence of sources, to the vacuum Bianchi differential equations:

$$\nabla_{[\alpha}\, W_{\beta\gamma]\delta\epsilon} = 0. \tag{293}$$

We can write these as

$$D\,W = 0 \tag{294}$$

to emphasize the analogy with the exterior derivative. These are the analogues of the Maxwell equations

$$dF = 0.$$

However, D is not an exterior differential operator, so $D^2 \neq 0$. The equation $D^2 W = 0$, a differential consequence of the vacuum Bianchi equations, is in fact the algebraic condition

$$R_{\mu}^{\ \alpha\beta\gamma}\, {}^*W_{\nu\alpha\beta\gamma} - R_{\nu}^{\ \alpha\beta\gamma}\, {}^*W_{\mu\alpha\beta\gamma} = 0. \tag{295}$$

Here, $R_{\alpha\beta\gamma\delta}$ is the curvature of the underlying metric $g_{\mu\nu}$. One can ask: What about the analogues of the other Maxwell equations

$$d\ F^* = 0 \quad \text{(in the absence of sources)?}$$

The remarkable fact is that the equations

$$D\ {}^*W = 0 \quad (D\ W^* = 0) \tag{296}$$

are *equivalent* to the equations $DW = 0$. In components, the equations $D\ {}^*W = 0$ read

$$\nabla^\alpha\ W_{\alpha\beta\gamma\delta} = 0. \tag{297}$$

We shall presently define an energy-momentum tensor analogous to that for Maxwell's equations:

$$Q_{\alpha\beta} = \frac{1}{2}\left(F_{\alpha\rho}\ F_\beta^{\ \rho} + F^*_{\alpha\rho}\ F^{*\ \rho}_\beta \right) \tag{298}$$

(on a 4-dimensional spacetime manifold). This tensor had already been discovered by L. Bel and I. Robinson in the case $W = R$ of a metric g satisfying the Einstein vacuum equations. We define

$$Q_{\alpha\beta\gamma\delta} = \frac{1}{2}\left(W_{\alpha\rho\gamma\sigma}\ W_{\beta\ \delta}^{\ \rho\ \sigma} + {}^*W_{\alpha\rho\gamma\sigma}\ {}^*W_{\beta\ \delta}^{\ \rho\ \sigma} \right). \tag{299}$$

We call this tensor the *Bel–Robinson tensor*. It is a *totally symmetric* quartic form in the tangent space at each point (a 4-covariant tensor field) which is *trace-free* with respect to any pair of indices.

Recall that the electromagnetic energy-momentum tensor satisfies the positivity condition

$$Q\ (X_1, X_2) \geq 0$$

for any pair X_1, X_2 of future-directed timelike vectors at a point with equality if and only if F vanishes at that point.

Similarly, the Bel–Robinson tensor has the property

$$Q\ (X_1,\ X_2,\ X_3,\ X_4) \geq 0 \tag{300}$$

for any quadruplet X_1, X_2, X_3, X_4 of vectors at a point all of which are future-directed timelike, with equality if and only if W vanishes at this point. The above are the algebraic properties of Q, all of which follow from the algebraic properties of W. Moreover, if W is a solution of the vacuum Bianchi equations, then Q is divergence-free:

$$\nabla^\alpha\ Q_{\alpha\beta\gamma\delta} = 0. \tag{301}$$

(This is analogous to the fact that, in electromagnetic theory, $dF = 0$ and $d^*F = 0$ imply $\nabla^\alpha Q_{\alpha\beta} = 0$.)

Suppose now that we are given three vector fields X, Y, Z all of which are future-directed, non-spacelike and generate conformal isometries of (M, g). We denote by

$$^{(U)}\hat{\pi} = \widehat{\mathcal{L}_U g} \tag{302}$$

the trace-free part of $\mathcal{L}_U g$ for any vector field U. If U generates conformal isometries, then

$$^{(U)}\hat{\pi} = 0. \tag{303}$$

We consider the vector field

$$P^\mu = -Q^\mu_{\ \alpha\beta\gamma}\, X^\alpha\, Y^\beta\, Z^\gamma.$$

Then, we have

$$\begin{aligned}
\nabla_\mu\, P^\mu &= -\underbrace{(\,\nabla_\mu\, Q^\mu_{\ \alpha\beta\gamma}\,)}_{=0}\, X^\alpha\, Y^\beta\, Z^\gamma \\
&\quad - Q^\mu_{\ \alpha\beta\gamma}\, (\nabla_\mu X^\alpha)\, Y^\beta\, Z^\gamma \\
&\quad - Q^\mu_{\ \alpha\beta\gamma}\, X^\alpha\, (\nabla_\mu\, Y^\beta)\, Z^\gamma \\
&\quad - Q^\mu_{\ \alpha\beta\gamma}\, X^\alpha\, Y^\beta\, (\nabla_\mu\, Z^\gamma).
\end{aligned} \tag{304}$$

We write

$$\begin{aligned}
Q^\mu_{\ \alpha\beta\gamma}\, (\,\nabla_\mu\, X^\alpha\,)\, Y^\beta\, Z^\gamma &= \frac{1}{2}\, Q^{\mu\alpha}_{\quad\beta\gamma}\, (\,\nabla_\mu\, X_\alpha + \nabla_\alpha\, X_\mu\,)\, Y^\beta\, Z^\gamma \\
&= \frac{1}{2}\, Q^{\mu\alpha}_{\quad\beta\gamma}\ {}^{(X)}\pi_{\mu\alpha}\, Y^\beta\, Z^\gamma \\
&= \frac{1}{2}\, Q^{\mu\alpha}_{\quad\beta\gamma}\ {}^{(X)}\hat{\pi}_{\mu\alpha}\, Y^\beta\, Z^\gamma
\end{aligned}$$

where we have used the symmetric and trace-free nature of Q, and similarly with X, Y, Z replaced by Y, Z, X and Z, X, Y respectively. Thus, recalling again the totally symmetric nature of Q, (304) becomes

$$\begin{aligned}
\nabla_\mu\, P^\mu &= -\frac{1}{2}\,\{\, Q^{\mu\alpha}_{\quad\beta\gamma}\ {}^{(X)}\hat{\pi}_{\mu\alpha}\, Y^\beta\, Z^\gamma \\
&\qquad + Q^{\mu\ \beta}_{\ \alpha\ \gamma}\, X^\alpha\ {}^{(Y)}\hat{\pi}_{\mu\beta}\, Z^\gamma \\
&\qquad + Q^{\mu\ \ \gamma}_{\ \alpha\beta}\, X^\alpha\, Y^\beta\ {}^{(Z)}\hat{\pi}_{\mu\gamma}\,\} \\
&= 0.
\end{aligned} \tag{305}$$

For, ${}^{(X)}\hat{\pi} = {}^{(Y)}\hat{\pi} = {}^{(Z)}\hat{\pi} = 0$, as X, Y, Z are by assumption conformal Killing fields. Defining then the corresponding 3-form ${}^*P$

$$ {}^*P_{\nu\kappa\lambda} = P^{\mu}\, \epsilon_{\mu\nu\kappa\lambda}, $$

the fact that P is divergence-free is equivalent to the fact that ${}^*P$ is closed:

$$ d\, {}^*P = 0. \tag{306} $$

We consider the following three cases.

Case 1. We integrate (306) over the domain bounded by the initial Cauchy hypersurface $\mathcal{H}_0$ and a Cauchy hypersurface $\mathcal{H}_t$ to the future of $\mathcal{H}_0$.

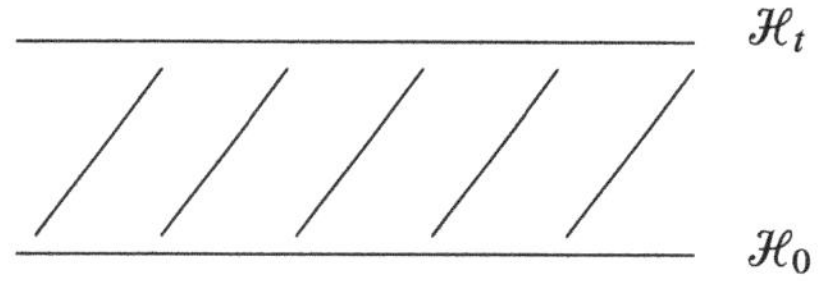

Figure 19

Case 2. We integrate (306) over the domain bounded by the initial Cauchy hypersurface $\mathcal{H}_0$ and an outgoing null hypersurface C_u.

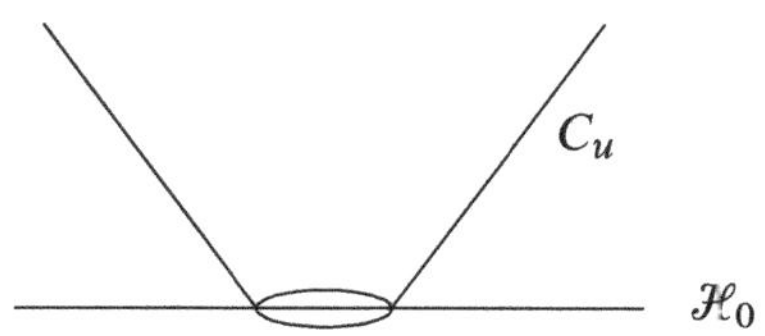

Figure 20

Case 3. We integrate (306) over the domain bounded by the initial Cauchy hypersurface $\mathcal{H}_0$ and an outgoing null hypersurface C_u capped in the past by a portion of the (complete) Cauchy hypersurface $\mathcal{H}_t$.

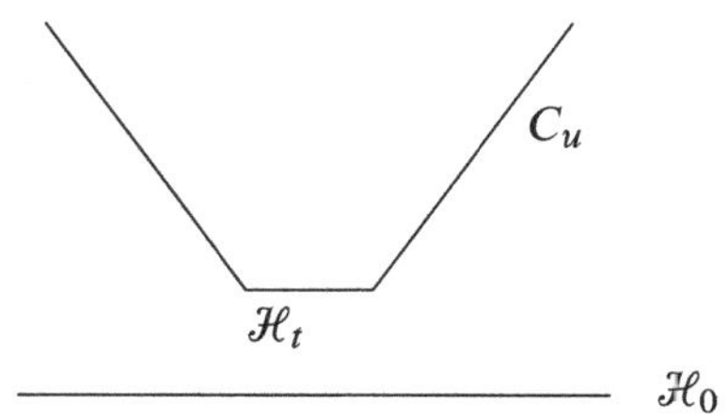

Figure 21

We then obtain

$$\int_{\mathcal{H}_t} {}^*P = \int_{\mathcal{H}_0} {}^*P \quad \text{in Case 1,} \tag{307}$$

$$\int_{C_u} {}^*P \leq \int_{\mathcal{H}_0} {}^*P \quad \text{in Case 2,} \tag{308}$$

$$\int_{C_u^c} {}^*P \leq \int_{\mathcal{H}_0} {}^*P \quad \text{in Case 3,} \tag{309}$$

where C_u^c is C_u capped by $\mathcal{H}_t$ in the past. Here, all quantities are non-negative.

In particular,

$$\int_{\mathcal{H}_t} {}^*P = \int_{\mathcal{H}_t} P^N \, d\mu_{\bar{g}_t} = \int_{\mathcal{H}_t} Q\,(N,\, X,\, Y\; Z)\, d\mu_{\bar{g}_t} \geq 0,$$

where N is the future-directed unit (timelike) normal to $\mathcal{H}_t$ and $d\mu_{\bar{g}_t}$ the volume element of the induced metric $\bar{g}_t$ on $\mathcal{H}_t$.

Now, given a Weyl field W and a vector field X, the Lie derivative with respect to X of W, that is $\mathcal{L}_X W$, is not in general a Weyl field, because it does not satisfy the vanishing trace condition. We can however define a modified Lie derivative $\hat{\mathcal{L}}_X W$ which is a Weyl field:

$$\begin{aligned} \hat{\mathcal{L}}_X\, W_{\alpha\beta\gamma\delta} = \mathcal{L}_X\, W_{\alpha\beta\gamma\delta} &- \frac{1}{2}\,(\,\hat{\pi}_\alpha^{\ \mu}\, W_{\mu\beta\gamma\delta} + \hat{\pi}_\beta^{\ \mu}\, W_{\alpha\mu\gamma\delta} \\ &+ \hat{\pi}_\gamma^{\ \mu}\, W_{\alpha\beta\mu\delta} + \hat{\pi}_\delta^{\ \mu}\, W_{\alpha\beta\gamma\mu}\,) \\ &- \frac{1}{8}\,\mathrm{tr}\,\pi\; W_{\alpha\beta\gamma\delta}. \end{aligned} \tag{310}$$

Here, as in the preceding, $\pi_{\alpha\beta} = \mathcal{L}_X g_{\alpha\beta}$ and $\hat{\pi}_{\alpha\beta}$ is the trace-free part of $\pi_{\alpha\beta}$.

The modified Lie derivative commutes with the Hodge dual,

$$\hat{\mathcal{L}}_X\, {}^*W = {}^*\hat{\mathcal{L}}_X\, W. \tag{311}$$

Conformal properties of the Bianchi equations. We consider next the conformal properties of the Bianchi equations.

Let f be a conformal isometry of the underlying manifold (M, g),

$$f^*g = \Omega^2\, g. \tag{312}$$

If W is a Weyl field satisfying $DW = 0$ on (M, g), then $\Omega^{-1} f^* W$ is also a Weyl field satisfying the same equation on (M, g). This follows from the fact, readily established by a straightforward calculation, that if W is a Weyl field satisfying the equation $DW = 0$ on (M, g), then, for any conformal factor Ω, $\tilde{W} = \Omega^{-1} W$ is also a Weyl field satisfying the equation $\tilde{D}\tilde{W} = 0$ on $(M, \tilde{g})$, where $\tilde{g} = \Omega^{-2} g$.

Remark. Recall that if W is the conformal curvature tensor of (M, g), then the conformal curvature tensor $\tilde{W}$ of $(M, \tilde{g})$ with $\tilde{g} = \Omega^{-2} g$ is

$$\tilde{W} = \Omega^{-2} W.$$

So the transformation $W \mapsto \Omega^{-1} W$ considered above is not related to this.

Suppose now that X is a vector field generating a 1-parameter group $\{f_t\}$ of conformal isometries of (M, g) (a conformal Killing field). Then if W is a solution of the Bianchi equations, so is $\Omega_t^{-1} f_t^* W$ for each t. By the linearity of the Bianchi equations

$$\frac{d}{dt} \Omega_t^{-1} f_t^* W \Big|_{t=0} = \hat{\mathcal{L}}_X W \tag{313}$$

is likewise a solution of the same equations. We see that the term $-\frac{1}{8} \operatorname{tr} \pi W$ in $\hat{\mathcal{L}}_X W$ comes from the conformal weight Ω^{-1}. Thus, if (M, g) possesses a non-trivial conformal isometry group, we can derive conserved quantities of arbitrary order by placing in the role of W the iterated (modified) Lie derivatives,

$$\hat{\mathcal{L}}_{X_{i_1}} \dots \hat{\mathcal{L}}_{X_{i_n}} W,$$

an n-th order Weyl field. Here, $i_1, \dots, i_n \in \{1, \dots, m\}$, with m being the dimension of the conformal group of (M, g) and $\{X_1, \dots, X_m\}$ being the generating conformal Killing fields.

4.2.3 Resolution of the second difficulty. We turn to the second difficulty, namely the fact that a general spacetime has only a trivial conformal isometry group.

The crucial observation here is that a spacetime which arises from arbitrary asymptotically flat initial data is itself expected to be asymptotically flat at spacelike infinity and at future null infinity in general, and also, under a suitable smallness restriction of the initial data, at timelike infinity as well.

We thus expect that, under the present circumstances, the spacetime approaches the Minkowski spacetime as the time tends to infinity. Now, the Minkowski spacetime possesses a large conformal isometry group. We thus expect to be able to define 'in the limit' $t \to \infty$ the action of a subgroup, at least, of the conformal group of Minkowski spacetime, as the action of a conformal isometry group in the limit $t \to \infty$.

Then the problem is to extend this action backwards in time up to the initial hypersurface in such a way as to obtain an action of the said subgroup globally, which is globally close to being the action of a conformal isometry group, in the sense that the deformation tensors $\hat{\pi}$ of the generating vector fields are globally small, and tend suitably fast to 0 as $t \to \infty$.

It turns out that we can only define the action of the subgroup of the conformal group of Minkowski spacetime corresponding to:

1. The time translations.
2. The scale transformations.
3. The inverted time translations.
4. The spatial rotation group O(3).

This is due to the fact that a non-trivial spacetime, corresponding to asymptotically flat initial data, has a non-zero total mass, therefore a non-zero energy-momentum vector (which can be considered to be a vector at the ideal point at spacelike infinity). Therefore an O(3) subgroup is singled out which leaves this vector invariant.

The group of time translations is the easiest to define. This corresponds to the choice of a canonical time function t. This is the canonical *maximal* time function relative to which the (spatial) linear momentum P^i vanishes. The generating vector field T has already been defined. The integral curves of T are the family of timelike curves orthogonal to the maximal hypersurfaces $\mathcal{H}_t$ and are parametrized by t. The corresponding group $\{f_\tau\}$ is such that f_τ is a diffeomorphism of $\mathcal{H}_t$ onto $\mathcal{H}_{t+\tau}$.

The rotation group O(3) is to satisfy the condition that it takes any given hypersurface $\mathcal{H}_t$ onto itself. To define the action of O(3) on $\mathcal{H}_t$ we must define the orbit of O(3) through a given point p. The construction is accomplished with the introduction of another function u, which is called an *'optical function'* as it is a solution of the *eikonal equation*

$$g^{\mu\nu}\,\partial_\mu\, u\;\partial_\nu\, u = 0. \tag{314}$$

This equation expresses the fact that the level sets C_u of u are null hypersurfaces. Then the 2-surfaces of intersection,

$$S_{t,u} = \mathcal{H}_t \cap C_u, \tag{315}$$

shall be the orbits of the rotation group O(3) on each $\mathcal{H}_t$. Moreover, the function u shall also be used to define the vector fields S and K generating the scale transformations and inverted time translations respectively.

Construction of the optical function *u*. Thus, the most essential step is the construction of the appropriate function u. The construction starts by choosing a 2-surface $S_{0,0}$, diffeomorphic to S^2, in the initial hypersurface $\mathcal{H}_0$. We consider $\partial J^+(S_{0,0})$, the boundary of the future of $S_{0,0}$, in the spacetime which is assumed to have been constructed. This has an outer as well as an inner component. The outer component is generated by the congruence of outgoing null geodesic normals to $S_{0,0}$ and the inner component is generated by the congruence of incoming null geodesic normals to $S_{0,0}$. We define the level set C_0 (the 0-level set of u) to be this outer component. It is an outgoing null hypersurface.

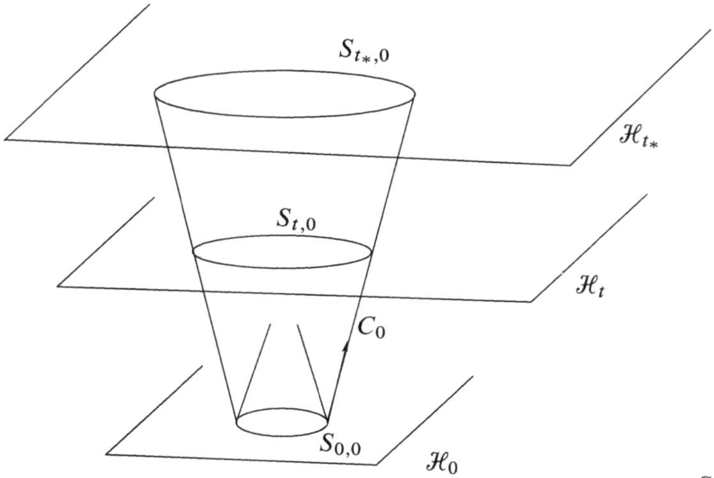

$S_{t,u} = \mathcal{H}_t \cap C_u$

Figure 22

Now, there is considerable freedom in the choice of $S_{0,0}$. However, the choice is subject to the condition that the null geodesic generators of C_0 have no future end-points.

We must now define the other level sets C_u with $u \neq 0$. These shall also be outgoing null hypersurfaces, thus u will, by construction, be a solution of the eikonal equation (314)

$$g^{\mu\nu}\,\partial_\mu\, u\,\partial_\nu\, u = 0.$$

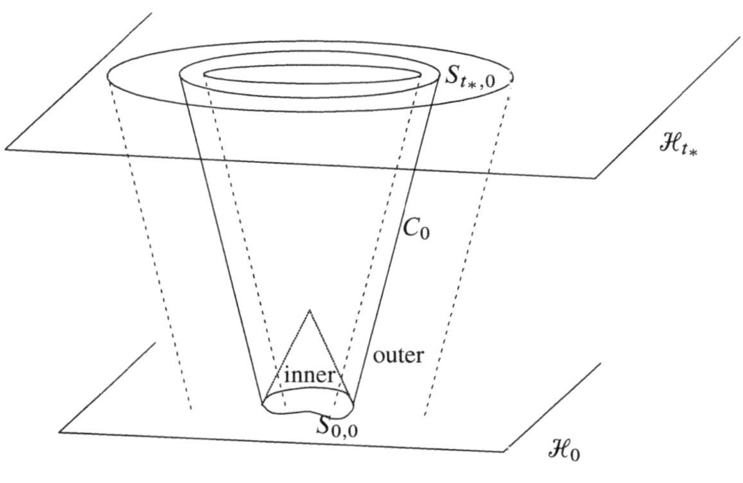

Figure 23

Consider the surfaces (315)

$$S_{t,u} = \mathcal{H}_t \cap C_u.$$

We want to impose the following condition: The surfaces $S_{t_*,u}$ in $\mathcal{H}_{t_*}$ must become equally spaced as $t_* \to \infty$. That is, $u|_{\mathcal{H}_{t_*}}$ must tend to minus the signed distance function from $S_{t_*,0}$ on $\mathcal{H}_{t_*}$ as $t_* \to \infty$.

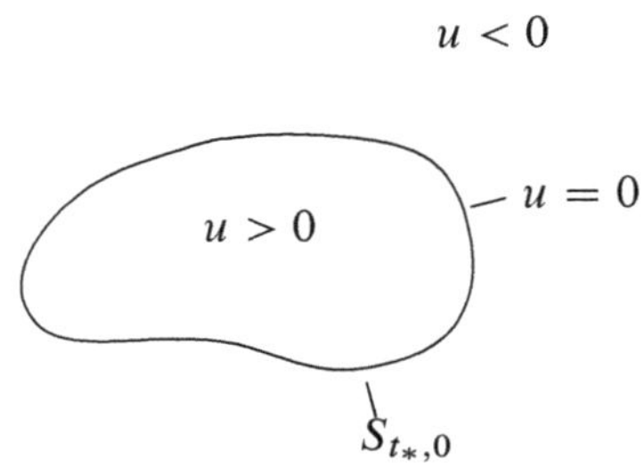

Figure 24

The global stability theorem is established by a continuity argument, in the course of which we are constructing a spacetime slab bounded in the future by the maximal hypersurface $\mathcal{H}_{t_*}$. The obvious choice of u on $\mathcal{H}_{t_*}$, namely minus the signed distance function from $S_{t_*,0}$, is *inappropriate*, because this distance function is only as smooth as the induced metric $\bar{g}_{t_*}$, not one order of differentiability better, which would be the maximal possible for a function on $(\mathcal{H}_{t_*}, \bar{g}_{t_*})$. This loss of one order of differentiability would result in failure of closure of the estimates. The continuity argument would then fail.

To overcome this difficulty, we define u on $\mathcal{H}_{t_*}$ in a different way, namely by solving a certain equation of motion of surfaces on $\mathcal{H}_{t_*}$, the initial surface being $S_{t_*,0}$. To keep the discussion simple, we neglect the second fundamental form k_{t_*} of $\mathcal{H}_{t_*}$, and consider the motion of a surface on a (3-dimensional) Riemannian manifold.

Given a function u on such a manifold, a function whose level sets define locally a foliation, we have the associated lapse function

$$a = (\bar{g}^{ij}\, \partial_i\, u\, \partial_j\, u)^{-\frac{1}{2}}, \tag{316}$$

which measures the normal separation of the leaves. We can think of a as the normal velocity of a surface, leaf of the foliation. An equation of motion of surfaces is then a rule which assigns a positive function a to a given surface. Given a surface S diffeomorphic to S^2, let f be the function

$$f = K - \frac{1}{4}\,(\mathrm{tr}\,\theta)^2, \tag{317}$$

where K is the Gauss curvature of S. (So, $f = 0$ for a round sphere in Euclidean space.) The rule which defines the equation of motion is

$$\Delta\!\!\!/ \log a = f - \bar{f}, \tag{318}$$

where $\Delta\!\!\!/$ is the Laplacian of the induced metric γ on S. In general, for any function f on S we denote by $\bar{f}$ the mean value of f on S. Equation (318) determines a up to a positive multiplicative constant. The freedom which is left corresponds to the freedom in relabeling the level sets of u. (We can remove the freedom by requiring $\overline{\log a} = 0$.)

To see what this equation of motion accomplishes, consider the trace of the 2nd variation equation for a 2-surface on a 3-dimensional Riemannian manifold, once the Gauss equation has been employed to express $\bar{R}_{33}$, e_3 being the unit outward normal to S, in terms of $\bar{R}$ and K. The Gauss equation reads

$$2K - (\operatorname{tr}\theta)^2 + |\theta|^2 = 2\bar{R} - \bar{R}_{33}.$$

Note also that since u is to decrease outwards, we have $e_3 = -\frac{1}{a}\frac{\partial}{\partial u}$. The general formula we then obtain is the following:

$$\frac{\partial \operatorname{tr}\theta}{\partial u} = \Delta\!\!\!/\, a + \frac{1}{2}\, a\, (\bar{R} + |\theta|^2 + (\operatorname{tr}\theta)^2 - 2K). \tag{319}$$

Now, neglecting the second fundamental form k of $\mathcal{H}_{t_*}$, the Gauss constraint equation of the imbedding of $\mathcal{H}_{t_*}$ in spacetime becomes simply

$$\bar{R} = 0. \tag{320}$$

It follows that if a is subject to equation (318) above, (319) reduces to the following propagation equation for $\operatorname{tr}\theta$:

$$\frac{1}{a}\,\frac{\partial \operatorname{tr}\theta}{\partial u} = \frac{1}{2}\,|\hat{\theta}|^2 + \frac{1}{2}\,(\operatorname{tr}\theta)^2 + |\nabla\!\!\!/ \log a|^2 - \bar{f}. \tag{321}$$

Here, $\hat{\theta}$ is the trace-free part of θ. All the curvature terms on the right-hand side have been eliminated. Moreover, by the Gauss–Bonnet theorem

$$\int_{S_\rho} K\, d\mu_\gamma = 4\pi,$$

hence

$$\bar{f} = \frac{1}{A}\int_{S_\rho} f\, d\mu_\gamma = \frac{4\pi}{A}\left(1 - \frac{1}{16\pi}\int_{S_\rho} (\operatorname{tr}\theta)^2\, d\mu_\gamma\right), \tag{322}$$

where A is the area of S_ρ. This propagation equation of $\operatorname{tr}\theta$ is to be considered in connection with the Codazzi equations

$$\nabla\!\!\!/^{B}\, \hat{\theta}_{AB} - \frac{1}{2}\,\nabla\!\!\!/_{A} \operatorname{tr}\theta = \bar{R}_{A3} \tag{323}$$

(complementing e_3 with $(e_A : A = 1,2)$ an arbitrary local frame field for S). These form an elliptic system on S for $\hat{\theta}$ given $\operatorname{tr}\theta$. Also, in estimating $\nabla\!\!\!/ \log a$ from the equation of motion, we appeal to the Gauss equation which expresses K as

$$K = \frac{1}{4}(\operatorname{tr}\theta)^2 - \frac{1}{2}|\hat{\theta}|^2 - \overline{R}_{33}. \tag{324}$$

Because of the fact that there are no curvature terms on the right-hand side of the propagation equation (321) and the fact that one order of differentiability is gained in inverting the 1st order elliptic operator in (323), we are able to obtain estimates for the second fundamental form θ of the level sets of u which are of one order of differentiability higher than the estimates for the curvature assumed. Then the level sets of u and u itself is shown to be three orders of differentiability smoother than the curvature or one order of differentiability smoother than the metric, as required.

The meaning of the equation of motion of surfaces. Here, we will consider the *Hawking mass* m (see [17]) of a surface S diffeomorphic to S^2 in a 3-dimensional Riemannian manifold of vanishing scalar curvature: $\overline{R} = 0$. (The energy of S is $4\pi m$.) We first define the *area-radius* r of S by

$$\text{Area}\,(S) = 4\pi r^2.$$

Definition 45. The Hawking mass m of a surface S diffeomorphic to S^2 in a 3-dimensional Riemannian manifold is

$$m = \frac{r}{2}\left(1 - \frac{1}{16\pi}\int_S (\operatorname{tr}\theta)^2 \, d\mu_\gamma\right). \tag{325}$$

If B is a small geodesic ball with center at a point p, then

$$\lim_{B \searrow p} \frac{m}{\text{Vol}(B)} = \frac{\overline{R}(p)}{16\pi}. \tag{326}$$

Recall that for the general (non-vacuum) Einstein equations,

$$\overline{R} = 4\,T^{00} \tag{327}$$

on a hypersurface with vanishing second fundamental form. We see that the limit at a point (326) captures only the energy density of matter. There is no gravitational contribution to the density at a point, the gravitational energy being of non-local character.

Proposition 9. *For suitable families of surfaces S whose interiors exhaust the 3-manifold we have*

$$m\,(S) \to M, \tag{328}$$

the total mass, provided the 3-manifold is strongly asymptotically Euclidean.

Proof. We are assuming that

$$\bar{g}_{ij} = \left(1 + \frac{M}{2\rho}\right)^4 \delta_{ij} + O_2\,(\rho^{-1-\epsilon}), \quad \epsilon > 0, \tag{329}$$

where $\rho = |x|$. The term $O_2(\rho^{-1-\epsilon})$ does not contribute to the limit of m. Thus the proposition follows if we show that the Hawking mass $m(S)$ of the coordinate sphere $S_\rho \to M$ as $\rho \to \infty$ in the case of the metric

$$\bar{g}_{ij} = \left(1 + \frac{M}{2\rho}\right)^4 \delta_{ij}, \tag{330}$$

the *Schwarzschild metric*. In fact, for any metric of the form

$$\bar{g} = \chi^4\, |dx|^2, \quad \chi = \chi(|x|),$$

changing to polar coordinates, we have

$$\bar{g} = \chi^4\,(\,d\rho^2 + \rho^2 \underbrace{\overset{\circ}{\gamma}_{AB}\,(y)\,dy^A dy^B}_{\text{standard metric on } S^2}), \quad \chi = \chi\,(\rho).$$

The arc length s along the rays from the origin is

$$s = \int_0^\rho \chi^2\,(\rho)\,d\rho.$$

The induced metric on S_ρ is $\gamma_{AB} = \rho^2 \chi^4\, \overset{\circ}{\gamma}_{AB}$. Then the second fundamental form θ_{AB} of S_ρ is given by

$$\begin{aligned} \theta_{AB} &= \frac{1}{2}\,\frac{\partial \gamma_{AB}}{\partial s} \\ &= \frac{1}{2\chi^2}\,\frac{\partial \gamma_{AB}}{\partial \rho} \\ &= \frac{1}{\chi^2}\left(\frac{1}{\rho} + \frac{2}{\chi}\frac{d\chi}{d\rho}\right)\gamma_{AB}. \end{aligned}$$

Hence

$$\operatorname{tr}\theta = \frac{2}{\rho\chi^2}\left(1 + \frac{2\rho}{\chi}\frac{d\chi}{d\rho}\right)$$

and $\sqrt{\det\gamma} = \rho^2\chi^4\sqrt{\det\overset{\circ}{\gamma}}$. Thus, we have

$$\int_S (\operatorname{tr}\theta)^2\, d\mu_\gamma = 16\,\pi\left(1 + \frac{2\rho}{\chi}\frac{d\chi}{d\rho}\right)^2.$$

Moreover, since $4\pi r^2 = \text{Area}(S_\rho) = \int_{S_\rho} d\mu_\gamma = 4\pi\rho^2\chi^4$ we have $r = \rho\chi^2$. We thus obtain

$$m\,(S_\rho) = \frac{1}{2}\,\rho\chi^2\left[1 - \left(1 + \frac{2\rho}{\chi}\frac{d\chi}{d\rho}\right)^2\right].$$

In the particular case $\chi = 1 + \frac{M}{2\rho}$ we find

$$m\,(S_\rho) = M.$$

Remark. 'Suitable' family of surfaces may be taken to mean the following:

$$\operatorname{tr}\theta = \frac{2}{r} + O(r^{-1-\epsilon}), \quad \epsilon > 0,\ r = \sqrt{\frac{A}{4\pi}},$$

and

$$|\hat{\theta}| = O(r^{-1-\epsilon}).$$

It is these two properties of the surfaces which are required in the above proof. □

Definition 46. Given an arbitrary local foliation with lapse function a, the mass aspect function of each leaf is

$$\mu = -\not\Delta \log a + K - \frac{1}{4}\,(\operatorname{tr}\theta)^2. \tag{331}$$

By the Gauss–Bonnet theorem, $\int_S K d\mu_\gamma = 4\pi$; hence we have

$$\begin{aligned}\int_S \mu\; d\mu_\gamma &= 4\pi\left(1 - \frac{1}{16\pi}\int_S (\operatorname{tr}\theta)^2\, d\mu_\gamma\right)\\ &= \frac{8\pi m}{r}.\end{aligned} \tag{332}$$

Thus

$$\bar{\mu} = \frac{2m}{r^3}. \tag{333}$$

Consider now the variation of m as we move through the foliation. We have

$$\frac{dm}{du} = \frac{1}{2}\frac{dr}{du}\frac{2m}{r} - \frac{r}{32\pi}\frac{d}{du}\int_S (\operatorname{tr}\theta)^2\, d\mu_\gamma. \tag{334}$$

Now, the general formula (319) reduces in the case $\overline{R} = 0$ under consideration to

$$\frac{\partial \operatorname{tr}\theta}{\partial u} = \not\Delta\; a + \frac{1}{2}\,a\,(\,|\theta|^2 + (\operatorname{tr}\theta)^2 - 2\,K).$$

We express this in terms of μ to obtain

$$\frac{1}{a}\frac{\partial \operatorname{tr}\theta}{\partial u} = \frac{1}{2}(\operatorname{tr}\theta)^2 + \frac{1}{2}|\hat{\theta}|^2 + |\nabla\!\!\!/ \log a|^2 - \mu. \tag{335}$$

Using the fact that

$$\frac{\partial d\mu_\gamma}{\partial u} = -a \operatorname{tr}\theta \, d\mu_\gamma,$$

we then deduce that

$$\frac{d}{du}\int (\operatorname{tr}\theta)^2 \, d\mu_\gamma = \int_S a \operatorname{tr}\theta \, \{\, |\hat{\theta}|^2 + 2\,|\nabla\!\!\!/ \log a|^2 - 2\,\mu \,\}\, d\mu_\gamma. \tag{336}$$

On the other hand, since

$$8\pi r \frac{dr}{du} = \frac{dA}{du} = -\int_S a \operatorname{tr}\theta \, d\mu_\gamma,$$

we obtain

$$\frac{dr}{du} = -\frac{r}{2}\,\overline{a \operatorname{tr}\theta}. \tag{337}$$

Now, in the formula for $\frac{d}{du}\int_S (\operatorname{tr}\theta)^2 d\mu_\gamma$ we have the term

$$-2\int_S a \operatorname{tr}\theta \, \mu \, d\mu_\gamma.$$

Writing

$$\mu = (\mu - \bar{\mu}) + \bar{\mu}$$

we have

$$-2\int_S a \operatorname{tr}\theta \, \bar{\mu}\, d\mu_\gamma = -8\pi r^2\, \overline{a \operatorname{tr}\theta}\, \bar{\mu} = -\frac{16\pi m}{r}\,\overline{a \operatorname{tr}\theta}, \tag{338}$$

by (333). Therefore, substituting (336) and (337) in (334), $-\frac{r}{32\pi}\cdot$ (338) cancels the first term on the right in (334). We thus remain with

$$\begin{aligned}\frac{dm}{du} = \frac{r}{16\pi}\Big\{ &-\int_S a \operatorname{tr}\theta \left(\frac{1}{2}|\hat{\theta}|^2 + |\nabla\!\!\!/ \log a|^2 \right) d\mu_\gamma \\ &+ \int_S (a \operatorname{tr}\theta - \overline{a \operatorname{tr}\theta})\,(\mu - \bar{\mu})\, d\mu_\gamma \Big\},\end{aligned} \tag{339}$$

where we have made use of the fact that $(\mu - \bar{\mu})$ has vanishing mean to rewrite

$$\int_S a \operatorname{tr}\theta \,(\mu - \bar{\mu})\, d\mu_\gamma \quad \text{as} \quad \int_S (a \operatorname{tr}\theta - \overline{a \operatorname{tr}\theta})\,(\mu - \bar{\mu})\, d\mu_\gamma.$$

The first integral in (339) is ≥ 0, provided that $\operatorname{tr}\theta \geq 0$. Thus, under this condition, the vanishing of the second integral implies that m is a non-increasing function of u. The second integral vanishes in the following two cases:

Case 1. $a \operatorname{tr}\theta = \overline{a \operatorname{tr}\theta}$: This is *inverse mean curvature flow.*

Case 2. $\mu = \bar{\mu}$: This is the *equation of motion of surfaces*, which we discussed above and which has the smoothing property required in the continuity argument.

Case 1 is a *parabolic equation.* The problem can be solved in this case in the *outward* direction only.

Case 2 may be thought of as an *ordinary differential equation* in the space of surfaces. For, the rule assigning the positive function a to a given surface S according to (318) assigns a function of the same differentiability class as the surface itself. This is because in inverting the 2nd order elliptic operator $\not\Delta$, two orders of differentiability are gained. The equation of motion of surfaces can be solved in *both* directions. We shall discuss general ordinary differential equations in the space of surfaces in a given 3-dimensional Riemannian manifold at the end of the present section.

The problem in Case 2 is actually solved outward globally (that is, for all $u \leq 0$) and inward up to a surface of area equal to a given fraction of the area of $S_0 (= S_{t_*,0})$. (Note that S_0 is the 0-level set of u.) It turns out that $\overline{R}_{33}$, $\overline{R}_{A3}$, $\overline{R}_{AB}$ have different decay properties. For fixed u and t_* large they decay like r^{-3}, r^{-2}, r^{-1}, respectively.

The proof of this semiglobal existence theorem for the equation of motion of surfaces is based on the hypothesis that there exists a background function u' with level sets $S'_{u'}$, such that $S'_0 = S_0$, and suitable assumptions hold on the geometric properties of this background foliation as well as on the components of $\overline{R}_{ij}$ in the decomposition with respect to the unit normal and the tangent plane to $S'_{u'}$. Once the surfaces $S_{t_*,u}$ have been constructed, we define C_u for $u \neq 0$ to be the *inner component of the causal past of* $S_{t_*,u}$. In this way the optical function u is constructed in the spacetime slab bounded by $\mathcal{H}_{t_*}$ and $\mathcal{H}_0$.

Next we define the vector fields S (scaling) and K (inverted time translation). The vector field T (time translation) has already been defined by the maximal foliation.

Consider a surface $S_{t,u} = \mathcal{H}_t \cap C_u$. At each point on $S_{t,u}$ we have two null normals L and $\underline{L}$, respectively outgoing and incoming, normalized by the condition that their components along T are equal to T. The integral curves of L are the null geodesic generators of the C_u, parametrized by t.

We define the function

$$\underline{u} = u + 2\,r,$$

where

$$r(t,u) = \sqrt{\frac{\operatorname{Area}(S_{t,u})}{4\pi}}.$$

Note that

$$T = \frac{1}{2}\,(\,L + \underline{L}\,).$$

Now let us define S and K according to

$$S = \frac{1}{2}\,(\,\underline{u}\,L + u\,\underline{L}\,),$$
$$K = \frac{1}{2}\,(\,\underline{u}^2\,L + u^2\,\underline{L}\,).$$

To define the action O(3) in the spacetime slab, we first consider the 'final' maximal hypersurface $\mathcal{H}_{t*}$. We consider on $\mathcal{H}_{t*}$ the vector field U:

$$U^i = a^2\,\bar{g}^{ij}\,\partial_j\,u.$$

The integral curves of U are orthogonal to the level sets of u on $\mathcal{H}_{t*}$, namely the surfaces $S_{t*,u}$ and are parametrized by u. (So, $Uu = 1$.) Let $\{\chi_\sigma\}$ be the 1-parameter group generated by U. Then χ_σ restricts to a diffeomorphism of $S_{t*,u}$ onto $S_{t*,u+\sigma}$. In particular,

$$\chi_u\colon S_{t*,0} \to S_{t*,u}$$

is a diffeomorphism. The pullback to $S_{t*,0}$ of the induced metric on $S_{t*,u}$ rescaled by r^{-2}, namely

$$\chi_u^*\,(r^{-2}\,\gamma)_{S_{t*,u}},$$

is shown to converge, as $u \to -\infty$ (that is, at spacelike infinity on $\mathcal{H}_{t*}$), to a metric $\overset{\circ}{\gamma}_{t*}$ of Gauss curvature equal to 1. Therefore, $(S_{t*,0}, \overset{\circ}{\gamma}_{t*})$ is isometric to the standard sphere. The rotation group O(3) then acts as the isometry group of $(S_{t*,0}, \overset{\circ}{\gamma}_{t*})$, the 'sphere at infinity'.

We then define the action of O(3) on $\mathcal{H}_{t*}$ by conjugation: Given a point $p \in S_{t*,u}$ and an element $O \in$ O(3), we consider the integral curve

$$\sigma \mapsto \chi_\sigma\,(p)$$

of U through p. As $\sigma \to -\infty$, this tends to a point q on the 'sphere at infinity'. In other words since $(S_{t*,0}, \overset{\circ}{\gamma}_{t*})$ is our model for the sphere at infinity, we can simply identify q with the point $\chi_{-u}(p) \in S_{t*,0}$. Then $Oq \in (S_{t*,0}, \overset{\circ}{\gamma}_{t*})$ is well defined. Finally, the point Op is the point $\chi_u(Oq) \in S_{t*,u}$. The vector fields $\Omega^{(a)}$ with $a = 1, 2, 3$ generating this action then satisfy

$$[U,\ \Omega^{(a)}] = 0,$$
$$[\,{}^{(a)}\Omega,\ {}^{(b)}\Omega] = \epsilon_{abc}\ \Omega^{(c)}$$

and are tangential to the surfaces $S_{t*,u}$. The last equations are the commutation relations of the Lie algebra of O(3).

This action of O(3) on $\mathcal{H}_{t_*}$ is then extended to the spacetime slab bounded by $\mathcal{H}_{t_*}$ and $\mathcal{H}_0$ by conjugation with the flow of L. The integral curves of L are the null geodesic generators of the hypersurfaces C_u and are parametrized by t. The 1-parameter group of diffeomorphisms generated by L maps the surfaces $S_{t,u}$ corresponding to the same value of u but different values of t onto each other. Given a point $p \in S_{t,u}$ and an element $O \in \mathrm{O}(3)$, we follow the integral curve of L through p at parameter value t to the point $p_* \in S_{t_*,u}$ at parameter value t_*. The action of O(3) on $\mathcal{H}_{t_*}$ defined above leads us to the point $Op_* \in S_{t_*,u}$. Finally, $Op \in S_{t,u}$ is defined to be the point at parameter value t along the integral curve of L through Op_* at parameter value t_*.

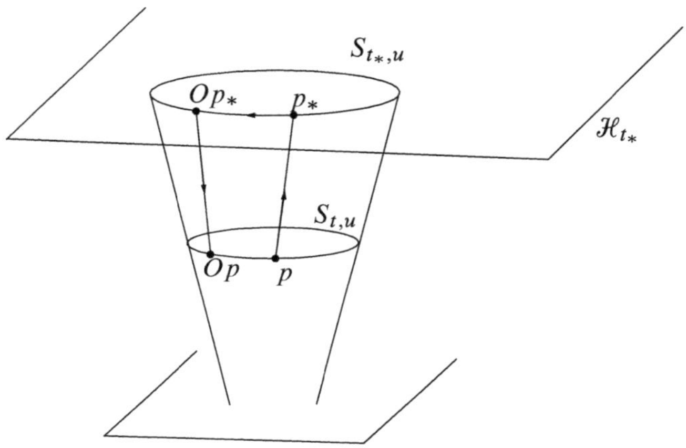

Figure 25

The $\Omega^{(a)}$ also satisfy

$$[L, \Omega^{(a)}] = 0,$$
$$[\Omega^{(a)}, \Omega^{(b)}] = \epsilon_{abc}\, \Omega^{(c)}$$

and are tangential to the surfaces $S_{t,u}$. Again, the last equations are the commutation relations of the Lie algebra of O(3).

General equations of motion of surfaces. We now give a general discussion of ordinary differential equations in the space of surfaces in a given Riemannian manifold (M, g), outlining how a local existence theorem for such equations is established. Let $\mathcal{A}$ be a rule which assigns to each surface S in M a positive function $\mathcal{A}(S)$ on S, of the same differentiability class as S. Then the problem we are considering is the following. Given an initial surface S_0 in M, find a function u defined in a neighborhood of S_0 in M, such that $u = 0$ on S_0 and for each level surface S_u of u,

$$|du|^{-1}\big|_{S_u} = \mathcal{A}(S_u). \tag{340}$$

Here du denotes the differential of u. To solve this problem we first consider the following simpler problem. Given an initial surface S_0 in M and a positive function a defined in a neighborhood of S_0 in M, find a function u defined in a smaller neighborhood of S_0 in M, such that $u = 0$ on S_0 and

$$|du|^{-1} = a \tag{341}$$

in this neighborhood. In other words, find locally a function u whose 0-level set is S_0 and whose associated lapse function is the given function a.

Equation (341) is a special case of the stationary *Hamilton–Jacobi equation*

$$H(d\phi) = E \tag{342}$$

for a function ϕ on a manifold M, the configuration space. Here H is the *Hamiltonian*,a function on T^*M, the phase space, and E is the energy constant. A particular class of Hamiltonians are Hamiltonians of the form

$$H(p) = \frac{1}{2}|p|^2 + V(q) \quad \forall p \in T_q^*M, \forall q \in M. \tag{343}$$

A Hamiltonian of this form describes the motion of a particle of mass 1 in a potential V in M. Equation (341) results if we set

$$\phi = u, \quad V = -\frac{1}{2}a^{-2}, \quad E = 0. \tag{344}$$

We shall presently discuss how solutions to the general stationary Hamilton–Jacobi equation (342) are constructed. We consider the *canonical equations* associated to the Hamiltonian H. Let $(q^1, \dots, q^n)$ be local coordinates on M. Then for q in the domain of this chart, we can expand $p \in T_q^*M$ as

$$p = p_i dq^i\big|_q.$$

The coefficients $(p_1, \dots, p_n)$ of the expansion constitute a system of linear coordinates for T_q^*M. Then $(q^1, \dots, q^n; p_1, \dots, p_n)$ are local coordinates on T^*M and the Hamiltonian is represented by a function of these. The canonical equations take in terms of such local coordinates the form

$$\frac{dq^i}{dt} = \frac{\partial H}{\partial p_i}, \quad \frac{dp_i}{dt} = -\frac{\partial H}{\partial q^i}; \quad i = 1, \dots, n. \tag{345}$$

Denoting by $t \mapsto (q(t), p(t))$ a solution of the canonical equations, a variation through solutions of a given solution, denoted by $t \mapsto (\dot q(t), \dot p(t))$, satisfies the equations of variation

$$\begin{aligned} \frac{d\dot q^i}{dt} &= \frac{\partial^2 H}{\partial p_i \partial q^j}\dot q^j + \frac{\partial^2 H}{\partial p_i \partial p_j}\dot p_j, \\ \frac{d\dot p_i}{dt} &= -\frac{\partial^2 H}{\partial q^i \partial q^j}\dot q^j - \frac{\partial^2 H}{\partial q^i \partial p_j}\dot p_j. \end{aligned} \tag{346}$$

The *canonical form* of T^*M is the 1-form θ on T^*M, given in the above local coordinates by

$$\theta = p_i dq^i. \tag{347}$$

Evaluating θ on a variation through solutions $(\dot{q}, \dot{p})$ of a given solution (q, p) we have

$$\theta \cdot (\dot{q}, \dot{p}) = p_i \dot{q}^i, \tag{348}$$

where the right-hand side is a function of t. A basic proposition of Classical Mechanics, which readily follows from the above equations, is

$$\frac{d}{dt}(\theta \cdot (\dot{q}, \dot{p})) = dL \cdot (\dot{q}, \dot{p}) \tag{349}$$

where L is the *Lagrangian*, which in the Hamiltonian picture is a function on T^*M, represented by

$$L = p_i \frac{\partial H}{\partial p_i} - H. \tag{350}$$

Also dL is the differential of L, thus

$$dL \cdot (\dot{q}, \dot{p}) = \frac{\partial L}{\partial q^i}\dot{q}^i + \frac{\partial L}{\partial p_i}\dot{p}_i. \tag{351}$$

Given now a closed surface S_0 in M, we construct a solution ϕ of (342) vanishing on S_0 as follows. To each point $q_0 \in S_0$ we associate a covector $p_0 \in T^*_{q_0}M$ which is required to vanish on $T_{q_0}S_0$ and satisfy $H(q_0, p_0) = E$. Let $t \mapsto (q(t; q_0), p(t; q_0))$ be the solution of the canonical equations (345) corresponding to the initial conditions (q_0, p_0). We then set along each solution trajectory

$$\phi(q(t; q_0)) = \int_0^t L(q(t'; q_0), p(t'; q_0))dt' + Et \quad \forall q_0 \in S_0. \tag{352}$$

This defines ϕ in a neighborhood of S_0 in M, and obviously ϕ vanishes on S_0. We shall presently show that ϕ satisfies (342). Consider a curve $\gamma\colon (-1, 1) \to S_0$ on S_0 through the point $q_0 : \gamma(0) = q_0$. Let $X \in T_{q_0}S_0$ be the tangent vector to this curve at $q_0 : X = \dot{\gamma}(0)$. Consider then the 1-parameter family of solutions of the canonical equations

$$\{t \mapsto (q(t; \gamma(s)), p(t; \gamma(s))) \ : \ s \in (-1, 1)\}.$$

The derivative with respect to s at $s = 0$ is a variation through solutions $t \mapsto (\dot{q}(t; (q_0, X)), \dot{p}(t; (q_0, X)))$ of the solution $t \mapsto (q(t; q_0), p(t; q_0))$. This variation is the solution of the equations of variation (346) corresponding to the initial conditions

$$\dot{q}(0; (q_0, X)) = X, \quad \dot{p}(0, (q_0, X)) = \frac{d}{ds} p_0(\gamma(s))\Big|_{s=0}.$$

Let us take the derivative of both sides of the equation

$$\phi(q(t;\gamma(s))) = \int_0^t L(q(t';\gamma(s)), p(t';\gamma(s)))dt' + Et \tag{353}$$

with respect to s at $s = 0$. We obtain, for the left-hand side,

$$\left.\frac{\partial \phi}{\partial q^i}\right|_{q(t;q_0)} \dot{q}^i(t;(q_0, X)),$$

and for the right-hand side,

$$\begin{aligned}
&\int_0^t \left\{ \frac{\partial L}{\partial q^i}\dot{q}^i(t';(q_0,X)) + \frac{\partial L}{\partial p_i}\dot{p}_i(t';(q_0,X)) \right\} dt' \\
&= \int_0^t \frac{d}{dt'}\left\{p_i(t';q_0)\dot{q}^i(t';(q_0,X))\right\} dt' \\
&= p_i(t;q_0)\dot{q}^i(t;(q_0,X))
\end{aligned}$$

by (349). Here we have taken account of the fact that

$$p_i(0;q_0)\dot{q}^i(0;(q_0,X)) = p_0 \cdot X = 0.$$

It follows that

$$\left(\left.\frac{\partial \phi}{\partial q^i}\right|_{q(t;q_0)} - p_i(t;q_0) \right) \dot{q}^i(t;(q_0,X)) = 0 \quad \forall X \in T_{q_0}S_0. \tag{354}$$

On the other hand, taking the derivative of both sides of (352) with respect to t we obtain, for the left-hand side,

$$\left.\frac{\partial \phi}{\partial q^i}\right|_{q(t;q_0)} \frac{dq^i}{dt}(t;q_0),$$

and for the right-hand side,

$$L(q(t;q_0), p(t;q_0)) + E = (L+H)(q(t;q_0), p(t;q_0)) = p_i(t;q_0)\frac{dq^i}{dt}(t;q_0)$$

by (350) and the first of the canonical equations (345). For, the canonical equations imply that the Hamiltonian is constant along trajectories, hence

$$H(q(t;q_0), p(t;q_0)) = E. \tag{355}$$

It follows that

$$\left(\left.\frac{\partial \phi}{\partial q^i}\right|_{q(t;q_0)} - p_i(t;q_0) \right) \frac{dq^i}{dt}(t;q_0) = 0. \tag{356}$$

Equations (354) and (356) together imply

$$\frac{\partial \phi}{\partial q^i} = p_i, \tag{357}$$

provided that the set of vectors

$$\{\dot{q}^i(t;(q_0, X)) \; : \; X \in T_{q_0} S_0\}$$

together with the vector

$$\frac{dq^i}{dt}(t;q_0)$$

span $T_{q(t;q_0)}M$. Moreover, in view of (355) and (357), ϕ is a solution of the stationary Hamilton–Jacobi equation (342).

We shall now show how the above is applied to construct a local solution of the problem associated to (340). We set up an iteration as follows. Given a positive function a_n defined in a neighborhood of the surface S_0 in M, we define the function u_n to be the solution of the stationary Hamilton–Jacobi equation

$$H_n(du_n) = 0, \tag{358}$$

which is negative in the exterior and positive in the interior of S_0. Here H_n is the Hamiltonian

$$H_n = \frac{1}{2}|p|^2 + V_n, \quad V_n = -\frac{1}{2}a_n^{-2}. \tag{359}$$

We then define the new positive function a_{n+1} by

$$a_{n+1}|_{S_n} = \mathcal{A}(S_n) \tag{360}$$

for each level surface S_n of the function u_n. The starting point of the iteration is the function $a_0 = 1$, in which case u_0 is the signed distance function from S_0 on M. The study of the convergence of this iteration then establishes a limit, $\lim_{n\to\infty} u_n = u$, which is a local solution of the equation of motion of surfaces (340).

4.2.4 The controlling quantity. Having defined the approximate conformal Killing fields, we consider the 1-form

$$P = P_0 + P_1 + P_2, \tag{361}$$

where

$$P_0 = -Q\,(R)\,(\cdot, \bar{K}, T, T), \tag{362}$$

$$P_1 = -Q\,(\hat{\mathfrak{L}}_O\, R)\,(\cdot, \bar{K}, \bar{K}, T) - Q\,(\hat{\mathfrak{L}}_T\, R)\,(\cdot, \bar{K}, \bar{K}, \bar{K}), \tag{363}$$

$$\begin{aligned} P_2 = &-Q\,(\hat{\mathfrak{L}}_O^2\, R)\,(\cdot, \bar{K}, \bar{K}, T) - Q\,(\hat{\mathfrak{L}}_O\, \hat{\mathfrak{L}}_T R)\,(\cdot, \bar{K}, \bar{K}, \bar{K}) \\ &- Q\,(\hat{\mathfrak{L}}_S\, \hat{\mathfrak{L}}_T R)\,(\cdot, \bar{K}, \bar{K}, \bar{K}) - Q\,(\hat{\mathfrak{L}}_T^2\, R)\,(\cdot, \bar{K}, \bar{K}, \bar{K}). \end{aligned} \tag{364}$$

Here, $\bar{K} = K + T$ is everywhere timelike future-directed (for $t \geq 0$). Also, $Q(W)$ is the Bel–Robinson tensor associated to the Weyl field W. Moreover, O stands for $\{{}^{(a)}\Omega;\ a = 1, 2, 3\}$, the generators of the action of O(3).

We define

$$E_1 = \sup_t \int_{\mathcal{H}_t} {}^*P, \tag{365}$$

$$E_2 = \sup_u \int_{C_u} {}^*P. \tag{366}$$

Then *the controlling quantity* is

$$E = \max\{E_1,\ E_2\}. \tag{367}$$

The quantities E_1 E_2 are defined in the spacetime slab $\mathcal{U}_{t_*} = \bigcup_{t\in[0,t_*]} \mathcal{H}_t$. So, E depends in fact on t_*.

Remark. P_1 and P_2 vanish for solutions which are invariant under rotations and time translations. They give us effective control on the solutions because of the following two facts:

1. The only spherically symmetric solution of the vacuum Einstein equations besides Minkowski spacetime is the Schwarzschild solution.
2. The only static solution of the vacuum Einstein equations besides Minkowski spacetime is the Schwarzschild solution.

Note that a static spacetime means a spacetime admitting a hypersurface orthogonal Killing field which is timelike at infinity. Here the Schwarzschild solution is excluded in view of the fact that the topology of the maximal hypersurfaces is $\mathbb{R}^3$. To control the spacetime curvature in terms of the quantity E, we consider separately the exterior region $\mathcal{E}$ and its complement, the interior region $\mathcal{I}$. The region $\mathcal{E}$ is defined by the inequality

$$\mathrm{Area}(S_{t,u}) \geq \theta\ \mathrm{Area}(S_{t,0}),$$

where θ is a constant, $0 < \theta < 1$.

Remark. The last term $Q(\hat{\mathcal{L}}_T^2 R)(\cdot, \bar{K}, \bar{K}, \bar{K})$ in P_2 is used in controlling the spacetime curvature in the 'wave zone', that is, in a neighbourhood of C_0 of the form $\bigcup_{u\in[-c,c]} C_u$ for a fixed positive constant c.

Remark. The term $Q(\hat{\mathcal{L}}_S\ \hat{\mathcal{L}}_T R)(\cdot, \bar{K}, \bar{K}, \bar{K})$ in P_2 is used to control the top order derivatives of the spacetime curvature in the interior region.

Note that the vector field S is timelike future-directed in the interior of C_0, it is null future-directed on C_0, and it is uniformly timelike in the region $\mathcal{I}$. We shall presently outline how control of the spacetime curvature in the interior region is achieved. We consider the decomposition of a Weyl field W with respect to a family of spacelike hypersurfaces with unit future-directed timelike normal e_0. Complementing e_0 with an arbitrary local frame field $(e_i : i = 1, 2, 3)$ for these hypersurfaces, we define:

Definition 47. The electric and magnetic parts of a Weyl field W with respect to a given family of spacelike hypersurfaces are the symmetric trace-free tensor fields E and H on these hypersurfaces given by

$$E_{ij} = W_{i0j0}, \tag{368}$$

$$H_{ij} = -{}^*W_{i0j0}. \tag{369}$$

Here we consider the electric-magnetic decomposition relative to the canonical maximal foliation $\{\mathcal{H}_t\}$. The *Bianchi equations for a Weyl field W in the presence of a source J* are the equations

$$D W = J. \tag{370}$$

The source J is called a *Weyl current*. Relative to an electric-magnetic decomposition these equations take the form

$$\operatorname{div} E = \rho_E, \tag{371}$$

$$\operatorname{curl} E + \hat{\mathcal{L}}_{e_0} H = \sigma_E, \tag{372}$$

$$\operatorname{div} H = \rho_H, \tag{373}$$

$$\operatorname{curl} H - \hat{\mathcal{L}}_{e_0} E = \sigma_H. \tag{374}$$

The right-hand sides here contain lower order terms involving the second fundamental form and lapse function of the foliation as well as the components of the current. Now if $\hat{\mathcal{L}}_S W$ is already controlled, we can decompose $\hat{\mathcal{L}}_{e_0} W$ into a term proportional to $\hat{\mathcal{L}}_S W$ which we can place on the right-hand side plus a term of the form $\hat{\mathcal{L}}_X W$, where X is a vector field tangential to the leaves of the foliation and satisfying

$$|X|_{\bar{g}} \le \theta < 1, \tag{375}$$

(θ a positive constant) in the interior region I. Then we obtain a system of the form

$$\operatorname{div} E = \rho'_E, \tag{376}$$

$$\operatorname{curl} E + \hat{\mathcal{L}}_X H = \sigma'_E, \tag{377}$$

$$\operatorname{div} H = \rho'_H, \tag{378}$$

$$\operatorname{curl} H - \hat{\mathcal{L}}_X E = \sigma'_H. \tag{379}$$

This system is uniformly elliptic in I by virtue of (375), allowing us to obtain interior estimates for E and H.

4.2.5 The continuity argument. Since the vector fields T, S, K, ${}^{(a)}\Omega$; $a = 1,2,3$, are not exact conformal Killing fields, $d\ {}^*P$ does *not* vanish. Thus, the integrals $\int_{\mathcal{H}_t} {}^*P$ and $\int_{C_u} {}^*P$ differ from an integral over $\mathcal{H}_0$ by error integrals which are spacetime integrals over the part of the spacetime slab bounded by $\mathcal{H}_t$ and $\mathcal{H}_0$, or C_u and $\mathcal{H}_0$, of expressions which are quadratic in the Weyl fields W and linear in the deformation tensors $\hat{\pi}$ of the vector fields. The point is to estimate these error integrals in terms of the controlling quantity E (see Step 2 below).

We introduce a certain set of assumptions on the main geometric properties of the two foliations, namely the $\{\mathcal{H}_t\}$ and $\{C_u\}$. These are called the *bootstrap assumptions*. They involve in particular

1. the quantities
$$\sup_{S_{t,u}}(r^2 K), \quad \inf_{S_{t,u}}(r^2 K),$$
where K is the Gauss curvature of $S_{t,u}$ and $4\pi r^2 = \text{Area}(S_{t,u})$;

2. the quantities
$$\sup_{S_{t,u}} a, \quad \inf_{S_{t,u}} a, \quad \sup_{S_{t,u}} \Phi, \quad \inf_{S_{t,u}} \Phi,$$
where a and Φ are the lapse functions of the two foliations
$$(a^{-2} = \bar{g}^{ij}\partial_i u \partial_j u, \quad \Phi^{-2} = -g^{\mu\nu}\partial_\mu t \partial_\nu t).$$

The isoperimetric constant of each $S_{t,u}$ depends on the quantities 1. The Sobolev inequalities on each $\mathcal{H}_t$ depend on these as well as the first of the quantities 2. The Sobolev inequalities of each C_u depend on the quantities 1 as well as the second of the quantities 2. The assumption is that the above quantities differ from their standard values (here: 1) by at most ϵ_0.

The above are the most important of the geometric quantities, as they control the Sobolev constants. There is a long list of additional assumptions involving the remaining geometric quantities, such as $\sup_{S_{t,u}}(\frac{r\,\mathrm{tr}\,\theta}{2})$ and $\inf_{S_{t,u}}(\frac{r\,\mathrm{tr}\,\theta}{2})$, where θ is the second fundamental form of $S_{t,u}$ relative to $\mathcal{H}_t$. All these are also to differ from their standard values by at most ϵ_0.

The continuity argument involves the following four steps. There is in addition a Step 0, which we shall discuss afterwards. We consider the maximal closed spacetime slab $\mathcal{U}_{t_*} = \bigcup_{t\in[0,t_*]} \mathcal{H}_t$ for which the bootstrap assumptions on the geometric properties of the two foliations $\{\mathcal{H}_t\}$ and $\{C_u\}$ hold with a constant ϵ_0.

Step 1 (Estimate of deformation tensors). We show that the bootstrap assumptions imply that the deformation tensors $\hat{\pi}$ of the fundamental vector fields T, S, K, ${}^{(a)}\Omega$ are bounded in $\mathcal{U}_{t_*}$, in appropriate norms, by another small constant ϵ_1 (depending continuously on ϵ_0 and tending to 0 as $\epsilon_0 \to 0$).

Step 2 (Error estimates). Using the result of Step 1 we estimate the error integrals as follows:

$$|\text{error integrals}| \leq C\,\epsilon_1\,E.$$

This yields the conclusion

$$E \leq C\,D + C\,\epsilon_1\,E.$$

Here, E is the controlling quantity and D is a quantity involving only the initial data. So, if ϵ_1 is suitably small, we have

$$E \leq C\,D.$$

Step 3. By analyzing the structure equations of the two foliations, we deduce that the geometric quantities entering the bootstrap assumptions are in fact bounded by CE. Therefore, by Step 2, under a suitable smallness restriction on D (the size of the initial data) we can conclude that the said geometric quantities are in fact bounded by $\frac{\epsilon_0}{2}$. Thus the inequalities in the bootstrap assumptions are not saturated up to time t_*.

Step 4. We extend the solution to the slab $\bigcup_{t\in[t_*,t_*+\delta]} \mathcal{H}_t$, for some suitably small $\delta > 0$. We first extend the optical function u_{t_*}, which was defined on the slab $\mathcal{U}_{t_*}$ (with final data on $\mathcal{H}_{t_*}$ the solution of the equation of motion of surfaces starting from $S_{t_*,0}$), by extending its level sets as null hypersurfaces, that is, by extending each null geodesic generator to the parameter interval $[t_*, t_* + \delta]$. We use this extension, which we denote by u'_{t_*}, or rather its restriction to $\mathcal{H}_{t_*+\delta}$, in the role of the background foliation, on the basis of which we construct new final data on $\mathcal{H}_{t_*+\delta}$ by solving the equation of motion of surfaces on $\mathcal{H}_{t_*+\delta}$ starting from the surface $S_{t_*+\delta,0}$. With this final data we then construct the new optical function $u_{t_*+\delta}$. We consider the geometric quantities associated to the maximal foliation $\{\mathcal{H}_t\}$ extended to the interval $[0, t_* + \delta]$ and to the null foliation $\{C_u\}$, where u is now $u_{t_*+\delta}$. By continuity, if δ is chosen suitably small, these quantities remain $\leq \epsilon_0$, contradicting the maximality of t_* unless of course $t_* = \infty$, in which case the theorem is proved.

We shall now discuss Step 0. This concerns the hypothesis on the initial data. Take a point $p \in \mathcal{H}_0$ and a positive real number λ (representing a length). Let d_p be the distance function on $\mathcal{H}_0$ from p. Setting

$$\begin{aligned} D\,(p,\,\lambda) = {} & \sup_{\mathcal{H}_0} \{\, \lambda^{-2}\,(d_p^2 + \lambda^2)^3\,|\,\overline{\mathrm{Ric}}\,|^2 \,\} \\ & + \lambda^{-3} \Bigg\{ \int_{\mathcal{H}_0} \sum_{l=0}^{3} (d_p^2 + \lambda^2)^{l+1}\,|\overline{\nabla}^l\,k|^2\,d\mu_{\bar g} \\ & + \int_{\mathcal{H}_0} \sum_{l=0}^{1} (d_p^2 + \lambda^2)^{l+3}\,|\overline{\nabla}^l\,B|^2\,d\mu_{\bar g} \Bigg\}, \end{aligned} \tag{380}$$

we define the invariant

$$D = \inf_{p \in \mathcal{H}_0, \lambda > 0} D(p, \lambda) \tag{381}$$

optimizing the choice of p and λ. This invariant represents the size of the initial data. In (380), B is the *Bach tensor*.

Definition 48. On a 3-dimensional Riemannian manifold the tensor field given by

$$B_{ij} = \epsilon_j^{\ ab} \, \overline{\nabla}_a \left(\overline{R}_{ib} - \frac{1}{4} \, \bar{g}_{ib} \, \overline{R} \right) \tag{382}$$

is called the *Bach tensor*.

Remark. We can write

$$\begin{aligned} B_{ij} &= \operatorname{curl} \hat{\overline{R}}_{ij} \\ &= \frac{1}{2} \left(\epsilon_i^{\ ab} \, \overline{\nabla}_a \, \hat{\overline{R}}_{bj} + \epsilon_j^{\ ab} \, \overline{\nabla}_a \, \hat{\overline{R}}_{bi} \right), \end{aligned} \tag{383}$$

where $\hat{\overline{R}}_{ij}$ is the trace-free part of the Ricci curvature of the 3-manifold, namely

$$\hat{\overline{R}}_{ij} = \overline{R}_{ij} - \frac{1}{3} \, \bar{g}_{ij} \, \overline{R}.$$

Thus, the Bach tensor is symmetric and trace-free.

Theorem 8 (Bach [2]). *The vanishing of the Bach tensor is necessary and sufficient for the* 3*-manifold to be locally conformally flat. That is,*

$$B_{ij} = 0 \iff \bar{g}_{ij} = \chi^4 \, e_{ij}, \tag{384}$$

where e_{ij} is flat, thus locally isometric to the Euclidean metric.

Recall the hypothesis that $(\mathcal{H}_0, \bar{g})$ be strongly asymptotically Euclidean:

$$\bar{g}_{ij} = \left(1 + \frac{M}{2|x|} \right)^4 \delta_{ij} + o\left(|x|^{-\frac{3}{2}}\right)$$

in an appropriate coordinate system in a neighbourhood of infinity. The principal part at spatial infinity is conformally flat, hence has vanishing contribution to the Bach tensor.

We now state Step 0.

Step 0. On the basis of the hypothesis that

$$\sup_{\mathcal{H}_0} \left\{ \lambda^{-2} \left(d_p^2 + \lambda^2\right)^3 |\overline{\mathrm{Ric}}|^2 \right\}$$

is suitably small for some $p \in \mathcal{H}_0$ and $\lambda > 0$, we show that $\exp_p$, the exponential mapping with base point p is a diffeomorphism of $T_p\mathcal{H}_0$ onto $\mathcal{H}_0$. We then have a foliation $\{S_\rho\}$ of $\mathcal{H}_0$ by the geodesic spheres with center p and radius ρ. We consider the following Hodge-type elliptic system for the traceless symmetric tensor field $\hat{\bar{R}}_{ij}$:

$$\operatorname{div}\widehat{\overline{\mathrm{Ric}}} = \frac{1}{6}\,\bar{d}\bar{R} = \frac{1}{6}\,\bar{d}\,(|k|^2), \tag{385}$$

$$\operatorname{curl}\widehat{\overline{\mathrm{Ric}}} = B, \tag{386}$$

$(\operatorname{tr}\widehat{\overline{\mathrm{Ric}}} = 0)$. Here, we denote by $\bar{d}\,f$ the differential of a function f on $\mathcal{H}_0$.

Remark. The equation (385) is simply the Bianchi identity:

$$\bar{\nabla}^j\,\bar{R}_{ij} - \frac{1}{2}\,\partial_i\,\bar{R} = 0.$$

Note that $D(p,\lambda)$ gives us control on k and B, thus on the right-hand sides of (385), (386). The theory of such elliptic systems then gives us estimates for the components of $\hat{\bar{R}}_{ij}$, hence for the components of $\bar{R}_{ij}$, relative to the foliation by the $\{S_\rho\}$. Denoting

$$\begin{aligned} \Pi^i{}_a\,\Pi^j{}_b\,\bar{R}_{ab} &= a_{ab},\\ \Pi^i{}_a\,\bar{R}_{ij}\,N^j &= b_a,\\ \bar{R}_{ij}\,N^i\,N^j &= c, \end{aligned}$$

we in fact obtain

$$\begin{aligned} \int_{\mathcal{H}_0}(d_p^2+\lambda^2)^2\,|\hat{a}|^2\,d\mu_{\bar{g}} &\le C\,D\,(p,\,\lambda),\\ \int_{\mathcal{H}_0}(d_p^2+\lambda^2)^2\,|b|^2\,d\mu_{\bar{g}} &\le C\,D\,(p,\,\lambda),\\ \int_{\mathcal{H}_0}(d_p^2+\lambda^2)^2\,|c-\bar{c}|^2\,d\mu_{\bar{g}} &\le C\,D\,(p,\,\lambda), \end{aligned}$$

where $\bar{c}$ is the mean value of c on S_ρ and $\hat{a}$ is the trace-free part of a. Moreover, we have

$$\operatorname{tr}a + c = \bar{R} = |k|^2$$

and

$$\bar{c}\rho^3 \to 2M \quad \text{as } \rho\to\infty\ (\rho = d_p|_{S_\rho}).$$

4.2.6 Estimates for the geometric quantities associated to the maximal foliation $\{\mathcal{H}_t\}$. We shall now sketch how Step 3 is accomplished. We begin with the recovery of the bootstrap assumptions concerning the maximal foliation $\{\mathcal{H}_t\}$. This is done by considering the *structure equations* of the foliation.

1. The *intrinsic* geometry of the $\mathcal{H}_t$ is controlled through the contracted Gauss equation
$$\bar{R}_{ij} - k_{im}\, k^m_j = E_{ij}. \tag{387}$$
Here, E_{ij} is the *electric* part of the Weyl curvature.
2. The *extrinsic* geometry of $\mathcal{H}_t$ is controlled through the uncontracted Codazzi equations
$$\bar{\nabla}_i\, k_{jm} - \bar{\nabla}_j\, k_{im} = \epsilon_{ij}^{\ \ n}\, H_{mn}. \tag{388}$$
Here, H_{ij} is the *magnetic* part of the Weyl curvature.
3. The lapse function Φ of the maximal foliation is controlled through the lapse equation
$$\bar{\triangle}\, \Phi - |k|^2\, \Phi = 0. \tag{389}$$

The contraction of the Codazzi equations gives the constraint equation
$$\bar{\nabla}^j\, k_{ij} - \partial_i \operatorname{tr} k = 0. \tag{390}$$

Here, $\operatorname{tr} k = 0$, consequently equations (388) are equivalent to the following Hodge-type system for the symmetric trace-free tensor field k:
$$\begin{aligned} \operatorname{div} k &= 0,\\ \operatorname{curl} k &= H. \end{aligned} \tag{391}$$

This is seen from the following remark.

Remark. Let S_{ij} be a symmetric 2-covariant tensor field on a 3-manifold $(\bar{M}, \bar{g}_{ij})$. Consider
$$\epsilon_i^{\ ab}\, \bar{\nabla}_a\, S_{bj} = C_{ij}.$$
Then the antisymmetric part
$$C_{ij} - C_{ji}$$
is equivalent to its dual
$$\frac{1}{2}\,(C_{ij} - C_{ji})\, \epsilon^{ij}_{\ \ m}.$$
This is equal to
$$\begin{aligned} \epsilon^{ij}_{\ \ m}\epsilon_i^{\ ab}\, \bar{\nabla}_a\, S_{bj} &= (\bar{g}^{ja}\, \delta_m^b - \bar{g}^{jb}\, \delta_m^a)\, \bar{\nabla}_a\, S_{bj}\\ &= \bar{\nabla}^j\, S_{mj} - \partial_m \operatorname{tr} S. \end{aligned}$$

On the other hand the symmetric part is

$$\frac{1}{2}\left(C_{ij} + C_{ji}\right) = (\operatorname{curl} S)_{ij}$$

by definition.

4.2.7 Estimates for the geometric quantities associated to the null foliation $\{C_u\}$.

We proceed to discuss the recovery of the bootstrap assumptions concerning the null foliation $\{C_u\}$.

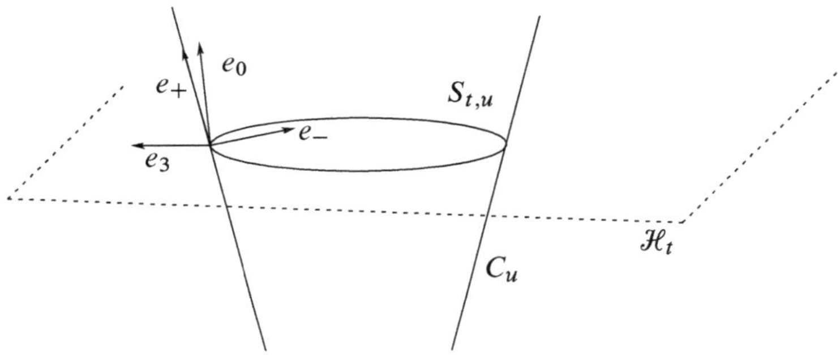

Figure 26

The geometry of a given null hypersurface C_u is described in terms of its sections $\{S_{t,u}\}$ by the maximal hypersurfaces $\mathcal{H}_t$. Let e_3 be the unit outward normal to $S_{t,u}$ in $\mathcal{H}_t$. Set

$$U = -a\, e_3. \tag{392}$$

The vector field U is characterized by the properties that it is tangential to the $\mathcal{H}_t$, orthogonal to the $\{S_{t,u}\}$ foliation of each $\mathcal{H}_t$ and satisfies

$$U\, u = 1.$$

Let $\{\chi_\sigma\}$ be the 1-parameter group of diffeomorphisms generated by U. Then χ_σ maps $S_{t,u}$ onto $S_{t,u+\sigma}$. Let e_0 be the unit future-directed timelike normal to $\mathcal{H}_t$. Then

$$T = \Phi\, e_0 \quad (T\, t = 1).$$

Let $\{\psi_\tau\}$ be the 1-parameter group of diffeomorphisms generated by L. Then ψ_τ maps $S_{t,u}$ onto $S_{t+\tau,u}$. We introduce the normalized null normals e_+, e_- to $S_{t,u}$ by

$$\begin{aligned} e_+ &= e_0 + e_3, \\ e_- &= e_0 - e_3. \end{aligned} \tag{393}$$

Then

$$\begin{aligned} L &= \Phi\, e_+ \quad (L\, t = 1), \\ \underline{L} &= \Phi\, e_- \quad (\underline{L}\, t = 1). \end{aligned}$$

The geometry of a given C_u is described by:

1. The intrinsic geometry of its sections $S_{t,u}$, that is, the induced metric γ and corresponding Gauss curvature K.
2. The second fundamental form χ of $S_{t,u}$ in C_u. This measures the deformation of $S_{t,u}$ under displacement along its outgoing null normal e_+, which is intrinsic to C_u. Completing e_+, e_- with e_A, $A = 1, 2$, an arbitrary local frame for $S_{t,u}$, χ is given by
$$\chi_{AB} = g\,(\nabla_{e_A}\, e_+,\, e_B). \tag{394}$$

The stacking of the C_u in the foliation $\{C_u\}$ is described by:

3. The function a, where
$$a^{-2} = \bar{g}^{ij}\, \partial_i\, u\, \partial_j\, u. \tag{395}$$

That is, a is the lapse function of the foliation of each $\mathcal{H}_t$ by the traces of the level sets of u. The deformation of $S_{t,u}$ under displacement along its incoming null normal e_-, which is transversal to C_u, is measured by $\underline{\chi}$, given by

$$\underline{\chi}_{AB} = g\,(\nabla_{e_A}\, e_-,\, e_B). \tag{396}$$

We have

$$\chi = \theta + \eta, \quad \underline{\chi} = -\theta\, - \eta, \tag{397}$$

where θ is the second fundamental form of $S_{t,u}$ relative to $\mathcal{H}_t$,

$$\theta_{AB} = \bar{g}\,(\bar{\nabla}_{e_A}\, e_3,\, e_B), \tag{398}$$

and η is the restriction of k, the second fundamental form of the maximal hypersurface $\mathcal{H}_t$ to $S_{t,u}$,

$$\eta_{AB} = g(\nabla_{e_A} e_0, e_B) = k(e_A, e_B). \tag{399}$$

The estimate of k_{ij}, therefore in particular of η_{AB}, has been discussed in the preceding section. The estimate of χ_{AB} shall be outlined below. The intrinsic geometry of $S_{t,u}$ is controlled by the equation for the Gauss curvature K:

$$K + \frac{1}{4}\,\mathrm{tr}\,\chi\,\mathrm{tr}\,\underline{\chi} - \frac{1}{2}\,\hat{\chi}\cdot\hat{\underline{\chi}} = -\rho \tag{400}$$

with $\rho = \frac{1}{4}R(e_-, e_+, e_-, e_+)$. Here $\hat{\chi}$ and $\hat{\underline{\chi}}$ are the trace-free parts of χ and $\underline{\chi}$ respectively.

We now outline the estimate of χ. Recall that C_u is defined to be the inner component of the boundary of the past of the surface $S_{t_*,u}$ on the final maximal hypersurface $\mathcal{H}_{t_*}$. Now, $\operatorname{tr}\chi$ satisfies the following propagation equation along each generator of C_u (parametrized by t):

$$\frac{1}{\Phi}\frac{\partial \operatorname{tr}\chi}{\partial t} = \nu \operatorname{tr}\chi - \frac{1}{2}(\operatorname{tr}\chi)^2 - |\hat{\chi}|^2. \tag{401}$$

Here

$$\nu = -\frac{1}{2}g(\nabla_{e_+}e_+, e_-), \quad \text{so} \quad \nabla_{e_+}e_+ = \nu e_+. \tag{402}$$

We have

$$\nu = \nabla_3 \log\Phi + \delta, \qquad \delta = k_{33}. \tag{403}$$

The propagation equation (401) is considered with a final condition on $\mathcal{H}_{t_*}$, namely $\operatorname{tr}\chi$ for the surface $\mathcal{S}_{t_*,u}$.

Important fact. *By virtue of the Einstein equations no curvature term appears on the right-hand side of* (401).

The propagation equation (401) is considered in conjunction with the null Codazzi equations:

$$\nabla\!\!\!/^{B}\, \hat{\chi}_{AB} - \frac{1}{2}\nabla\!\!\!/_{A} \operatorname{tr}\chi = \epsilon^{B}\, \hat{\chi}_{AB} - \frac{1}{2}\,\epsilon_A \operatorname{tr}\chi - \beta_A. \tag{404}$$

Here $\epsilon_A = k_{A3}$ and $\beta_A = \frac{1}{2}R(e_A, e_+, e_-, e_+)$. The equations (404) constitute an elliptic system for $\hat{\chi}$, given $\operatorname{tr}\chi$. Because of the fact that there are no curvature terms on the right-hand side of the propagation equation for $\operatorname{tr}\chi$ and the fact that one order of differentiability is gained in inverting the 1st order elliptic operator ($\operatorname{di\!\!/v}$ acting on trace-free symmetric 2-covariant $S_{t,u}$ tensor fields) in (404), we are able to obtain estimates for χ which are of one order of differentiability higher than the estimates for the spacetime curvature assumed.

To estimate a we consider the $S_{t,u}$-tangential 1-form

$$\zeta = \nabla\!\!\!/ \log a - \epsilon. \tag{405}$$

We have

$$\zeta_A = \frac{1}{2}\, g\,(\nabla_{e_-}e_+, e_A). \tag{406}$$

We also define

$$\lambda_A = g\,(\nabla_{e_+}e_-, e_A). \tag{407}$$

We have

$$\lambda = \nabla\!\!\!/ \log\Phi + \epsilon. \tag{408}$$

Now, with Π the projection to $S_{t,u}$, it holds that

$$\Pi\,[e_+,\, e_-] = -2\,(\zeta - \lambda). \tag{409}$$

Thus $(\zeta - \lambda)$ is the obstruction to integrability of the distribution of orthogonal timelike planes

$$\{\, (T_p\, S_{t,u}) \perp :\ p \in M \,\}. \tag{410}$$

This manifests itself in the non-commutativity of the 1-parameter groups generated by L and U:

$$[L,\, U] = \Phi\, a\, (\zeta - \lambda) \neq 0. \tag{411}$$

We consider the *mass aspect function*, defined in general by

$$\mu = -\not{\mathrm{div}}\, \zeta + K + \frac{1}{4} \operatorname{tr} \chi \operatorname{tr} \underline{\chi}. \tag{412}$$

This reduces to the expression (331) when $k_{ij} = 0$. The *Hawking mass* is defined in general by

$$m = \frac{r}{2}\left(1 + \frac{1}{16\pi} \int_S \operatorname{tr} \chi \operatorname{tr} \underline{\chi}\, d\mu_\gamma\right) \tag{413}$$

and we have

$$\bar{\mu} = \frac{2m}{r^3}. \tag{414}$$

The mass aspect function μ satisfies along the generators of C_u the following propagation equation:

$$\begin{aligned} \frac{1}{\Phi} \frac{\partial \mu}{\partial t} + \mu \operatorname{tr} \chi = {} & 2\, \hat{\chi} \cdot (\not{\nabla}\, \hat{\otimes}\, \zeta) - 2\, \zeta \cdot \beta \\ & - \frac{1}{2} \operatorname{tr} \chi\, (\not{\mathrm{div}}\, \lambda + |\lambda|^2 - \frac{1}{2}\, \hat{\chi} \cdot \hat{\underline{\chi}} - \rho) \\ & + (\zeta - \lambda) \cdot (\not{\nabla} \operatorname{tr} \chi - \epsilon \operatorname{tr} \chi) - \frac{1}{4} \operatorname{tr} \underline{\chi}\, |\hat{\chi}|^2 \\ & + \zeta \cdot \hat{\chi} \cdot \lambda. \end{aligned} \tag{415}$$

Here,

$$(\not{\nabla}\, \hat{\otimes}\, \zeta)_{AB} = \frac{1}{2}\, (\not{\nabla}_A\, \zeta_B + \not{\nabla}_B\, \zeta_A - \gamma_{AB}\, \not{\mathrm{div}}\, \zeta)$$

is the trace-free part of the symmetrized covariant derivative of ζ in $S_{t,u}$.

Important fact. *By virtue of the Einstein equations the right-hand side of* (415) *does not contain derivatives of the curvature.*

The propagation equation (415) with final data on $\mathcal{H}_{t*}$, the mass aspect function of the surface $S_{t*,u}$, is considered in conjunction with the following elliptic equation for a on $S_{t,u}$, which is simply a re-writing of the definition of μ:

$$\not\Delta \log a = -\mu + \not{\mathrm{div}}\, \epsilon + K + \frac{1}{4} \operatorname{tr} \chi \operatorname{tr} \underline{\chi}. \tag{416}$$

Because of the fact that there are no curvature terms on the right-hand side of the propagation equation for μ and the fact that two orders of differentiability are gained in inverting the 2nd order elliptic operator $\not\Delta$ in (416), we are able to obtain estimates for a which are of two orders of differentiability higher than the estimates for the spacetime curvature assumed.

The equation of motion of surfaces on the final maximal hypersurface $\mathcal{H}_{t*}$, takes in the general case the same form as in the special case where $k_{ij} = 0$, namely

$$\mu = \bar{\mu}; \tag{417}$$

however μ is now given by formula (412).

Remarks. $\eta_{AB} = k_{AB}$ is the leading part of k_{ij}, namely the part having the slowest decay. In fact, η can only be bounded by:

$$|\hat{\eta}| \leq C\ \{ \epsilon_0\, \tau_+^{-1}\, \tau_-^{-\frac{3}{2}} + \epsilon_0^2\, \tau_+^{-2} \}. \tag{418}$$

The weight functions τ_+, τ_-, are defined by

$$\tau_+ = \sqrt{1 + \underline{u}^2}, \quad \tau_- = \sqrt{1 + u^2}. \tag{419}$$

Note that

$$\operatorname{tr} \eta + \delta = \operatorname{tr} k = 0.$$

For the remaining components of k_{ij} we derive the bounds

$$|\epsilon| \leq C\ \{ \epsilon_0\, \tau_+^{-2}\, \tau_-^{-\frac{1}{2}} + \epsilon_0^2\, \tau_+^{-2} \}, \tag{420}$$

$$|\delta| \leq C\ \{ \epsilon_0\, \tau_+^{-\frac{5}{2}} + \epsilon_0^2\, \tau_+^{-2} \}. \tag{421}$$

On the other hand, we derive for χ the bounds

$$\begin{aligned} |\hat{\chi}| &\leq C\ \epsilon_0\, r^{-2}, \\ \left| \frac{r}{2} \operatorname{tr} \chi - 1 \right| &\leq C\ \epsilon_0\, r^{-1}. \end{aligned} \tag{422}$$

Consider the behavior on a given C_u as $t \to \infty$. Then

$$r\hat{\eta} \quad \text{tends to a non-trivial limit,}$$

while

$$r\,\delta \to 0.$$

Since we have

$$\chi = \theta + \eta,$$
$$\underline{\chi} = -\theta + \eta,$$

it follows that

$$\frac{r}{2}\,\mathrm{tr}\,\theta \to 1,$$

while

$$r\,\hat{\theta} \to -r\,\hat{\eta} \quad \text{tends to a non-trivial limit,}$$
$$r\,\hat{\underline{\chi}} \to 2\,r\,\hat{\eta} \quad \text{tends to a non-trivial limit.}$$

In particular, $|\hat{\theta}|^2/(\mathrm{tr}\,\theta)^2$ tends to a non-trivial limit. Consequently, the surfaces $S_{t,u}$ on a given C_u do not become umbilical as $t \to \infty$. In fact, the last limit is proportional to the amount of energy radiated per unit time per unit solid angle at a given retarded time and a given direction.

4.2.8 Decomposition of a Weyl field with respect to the surfaces $S_{t,u}$. In concluding our sketch of the proof of the global stability of Minkowski spacetime, we shall show in detail some of the more delicate estimates in Step 2. The discussion shall make use of the decomposition of a Weyl field and its associated Bel–Robinson tensor with respect to the surfaces $S_{t,u}$.

Consider the null frame e_+, e_- supplemented by e_A, $A = 1, 2$, a local frame field for $S_{t,u}$. The components of a Weyl field W in such a frame are

$$\alpha_{AB} = W\,(e_A,\, e_+,\, e_B,\, e_+), \qquad \underline{\alpha}_{AB} = W(e_A,\, e_-,\, e_A,\, e_-),$$
$$\beta_A = \frac{1}{2} W(e_A,\, e_+,\, e_-,\, e_+), \qquad \underline{\beta}_A = \frac{1}{2}\, W\,(e_A,\, e_-,\, e_-,\, e_+),$$
$$\rho = \frac{1}{4} W\,(e_-,\, e_+,\, e_-,\, e_+), \qquad \sigma\epsilon(e_A,\, e_B) = \frac{1}{2}\, W\,(e_A,\, e_B,\, e_-,\, e_+).$$

$\alpha, \underline{\alpha}$ are symmetric trace-free 2-covariant tensor fields on $S_{t,u}$, $\beta, \underline{\beta}$ are 1-forms on $S_{t,u}$, ρ, σ are functions on $S_{t,u}$.

Here, ϵ is the area 2-form of $S_{t,u}$.

Each of α, $\underline{\alpha}$ has two algebraically independent components,

each of β, $\underline{\beta}$ has two components,

and ρ, σ are two functions.

So, there are ten component-functions in all.

The components of $Q(W)$, the Bel–Robinson tensor associated to W, in the (timelike) plane spanned by e_+, e_-, are:

$$Q\,(W)\,(e_-,\, e_-,\, e_-,\, e_-) = 2\,|\underline{\alpha}|^2, \tag{423}$$

$$Q\,(W)\,(e_+,\, e_-,\, e_-,\, e_-) = 4\,|\underline{\beta}|^2, \tag{424}$$

$$Q\,(W)\,(e_+,\, e_+,\, e_-,\, e_-) = 4\,(\rho^2 + \sigma^2), \tag{425}$$

$$Q\,(W)\,(e_+,\, e_+,\, e_+,\, e_-) = 4\,|\beta|^2, \tag{426}$$

$$Q\,(W)\,(e_+,\, e_+,\, e_+,\, e_+) = 2\,|\alpha|^2. \tag{427}$$

Note that $e_0 = \frac{1}{2}(e_+ + e_-)$ is the unit future-directed normal to $\mathcal{H}_t$, and that

$$T = \Phi\, e_0 = \frac{1}{2}\,\Phi\,(e_+ + e_-), \tag{428}$$

$$\bar{K} = K + T = \frac{1}{2}\,\Phi\,(\tau_+^2\, e_+ + \tau_-^2 e_-), \tag{429}$$

the weight functions τ_+, τ_- being given by (419).

Recall now the controlling quantity $E = \max\{E_1, E_2\}$, where E_1 is defined in (365) as the supremum over t of an integral on $\mathcal{H}_t$ and E_2 in (366) as the supremum over u of an integral on C_u.

In what follows we shall use the following notation. Let f, g be positive functions. Then $f \sim g$ denotes that there exists a constant $C > 0$ such that:

$$C^{-1}\, f \le g \le C\, f.$$

Let us consider the integrants in E_1 and the ones in E_2. The integrants in E_1 are

$$Q\,(W)(\bar{K}, \bar{K}, T, e_0) \quad \text{for } W = \hat{\mathfrak{L}}_O R, \hat{\mathfrak{L}}_O^2 R,$$

$$Q\,(W)(\bar{K}, \bar{K}, \bar{K}, e_0) \quad \text{for } W = \hat{\mathfrak{L}}_T R,\; \hat{\mathfrak{L}}_O \hat{\mathfrak{L}}_T R, \hat{\mathfrak{L}}_S \hat{\mathfrak{L}}_T R$$

and we have

$$Q(W)(\bar{K}, \bar{K}, T, e_0) \sim \tau_-^4 |\underline{\alpha}|^2 + \tau_-^2 \tau_+^2 |\underline{\beta}|^2 + \tau_+^4 (\rho^2 + \sigma^2 + |\beta|^2 + |\alpha|^2) \tag{430}$$

$$Q(W)(\bar{K}, \bar{K}, \bar{K}, e_0) \sim \tau_-^6 |\underline{\alpha}|^2 + \tau_-^4 \tau_+^2\, |\underline{\beta}|^2 + \tau_-^2 \tau_+^4\, (\rho^2 + \sigma^2) + \tau_+^6 (|\beta|^2 + |\alpha|^2). \tag{431}$$

The integrants in E_2 are

$$Q\,(W)\,(\bar{K},\, \bar{K},\, T,\, e_+) \quad \text{for } W = \hat{\mathfrak{L}}_O R,\; \hat{\mathfrak{L}}_O^2 R,$$

$$Q\,(W)\,(\bar{K},\, \bar{K},\, \bar{K},\, e_+) \quad \text{for } W = \hat{\mathfrak{L}}_T R,\; \hat{\mathfrak{L}}_O \hat{\mathfrak{L}}_T R,\; \hat{\mathfrak{L}}_S \hat{\mathfrak{L}}_T R$$

and we have

$$Q(W)(\bar{K}, \bar{K}, T, e_+) \sim \tau_-^4 |\underline{\beta}|^2 + \tau_-^2 \tau_+^2 (\rho^2 + \sigma^2) + \tau_+^4 (|\beta|^2 + |\alpha|^2), \tag{432}$$

$$Q(W)(\bar{K}, \bar{K}, \bar{K}, e_+) \sim \tau_-^6 |\underline{\beta}|^2 + \tau_-^4 \tau_+^2 (\rho^2 + \sigma^2) + \tau_-^2 \tau_+^4 |\beta|^2 + \tau_+^6 |\alpha|^2. \tag{433}$$

4.2.9 The borderline error integrals. We now show how some of the error integrals in Step 2, requiring more careful treatment, are handled. These are the following:

1. $\displaystyle\int_{\mathcal{U}_{t*}} Q\,(W = \hat{\mathcal{L}}_O R \text{ or } \hat{\mathcal{L}}_O^2 R)_{\alpha\beta\gamma\delta}\ {}^{(\bar{K})}\hat{\pi}^{\alpha\beta}\ \bar{K}^\gamma\ T^\delta\ d\mu_g,$
2. $\displaystyle\int_{\mathcal{U}_{t*}} Q\,(W = \hat{\mathcal{L}}_O R \text{ or } \hat{\mathcal{L}}_O^2 R)_{\alpha\beta\gamma\delta}\ {}^{(T)}\hat{\pi}^{\alpha\beta}\ \bar{K}^\gamma\ \bar{K}^\delta\ d\mu_g,$
3. $\displaystyle\int_{\mathcal{U}_{t*}} Q\,(W = \hat{\mathcal{L}}_T R \text{ or } \hat{\mathcal{L}}_O \hat{\mathcal{L}}_T R \text{ or } \hat{\mathcal{L}}_S \hat{\mathcal{L}}_T R)_{\alpha\beta\gamma\delta}\ {}^{(\bar{K})}\hat{\pi}^{\alpha\beta}\ \bar{K}^\gamma\ \bar{K}^\delta\ d\mu_g.$

For the error integrals 1 the worst term is

$$\int_{\mathcal{U}_{t*}} Q\,(W)_{AB\gamma\delta}\ {}^{(\bar{K})}\hat{\pi}^{AB}\ \bar{K}^\gamma\ T^\delta\ d\mu_g$$

because ${}^{(\bar{K})}\hat{\pi}_{AB}$ contains $\tau_+^2\hat{\chi}_{AB}$ which is the part with the slowest decay. In fact this is merely bounded pointwise by $C\epsilon_1$. Moreover, the leading part of the integrant is obtained by taking the part of $\bar{K}$ with the largest weight, namely $\frac{1}{2}\Phi\tau_+^2 e_+$, and the $\frac{1}{2}\Phi e_-$ part of T:

$$\int_{\mathcal{U}_{t*}} |Q^{(\bar{K})}_{AB+-}\hat{\pi}^{AB}|\,\tau_+^2\ d\mu_g \le C\ \epsilon_1 \int_{\mathcal{U}_{t*}} |Q_{AB+-}|\,\tau_+^2\ d\mu_g.$$

Now, we have

$$Q_{AB+-}\ (W) = \{\,-2\,\beta\ \hat{\otimes}\ \underline{\beta} + 2\,\gamma\ (\rho^2 + \sigma^2)\,\}_{AB}.$$

Here, we denote

$$(x\hat{\otimes}y)_{AB} = x_A y_B + y_A x_B - \gamma_{AB}(x\cdot y).$$

We estimate

$$\begin{aligned}
&\int_{\mathcal{U}_{t*}} |\beta\ \hat{\otimes}\ \underline{\beta}|\ \tau_+^2\ d\mu_g \\
&\quad \le C \int du \left\{ \tau_-^{-2} \left(\int_{C_u} \tau_-^4\ |\underline{\beta}|^2\right)^{\frac{1}{2}} \cdot \left(\int_{C_u} \tau_+^4\ |\beta|^2\right)^{\frac{1}{2}} \right\} \\
&\quad \le C\ E \int \tau_-^{-2}\ du = C\ E,
\end{aligned}$$

comparing with (432) and noting that $\int_{-\infty}^{+\infty} \tau_-^{-2}\ du = \pi$. Since the geometric mean of the weights of $\beta,\ \underline{\beta}$ in (432) is the weight of $(\rho,\ \sigma)$, the term in $(\rho,\ \sigma)$ in Q_{AB+-} can be estimated in the same way. We thus obtain

$$|\,\text{error integrals 1}\,| \le C\ \epsilon_1\ E.$$

For the error integrals 2 the worst term is

$$\int_{\mathcal{U}_{t*}} Q\,(W)_{AB\gamma\delta}\ {}^{(T)}\hat{\pi}^{AB}\ \bar{K}^{\gamma}\bar{K}^{\delta}\ d\mu_g$$

because ${}^{(T)}\hat{\pi}_{AB}$ contains $\hat{\eta}_{AB}$ which decays pointwise only like $\tau_+^{-1}\tau_-^{-1}$. The leading part of this comes from the part of $\bar{K}^{\gamma}\bar{K}^{\delta}$ of the largest weight, namely from $(\frac{1}{2}\Phi)^2\tau_+^4 e_+^{\gamma} e_+^{\delta}$:

$$\int_{\mathcal{U}_{t*}} |Q_{AB++}\ {}^{(T)}\hat{\pi}^{AB}|\ \tau_+^4\ d\mu_g \leq C\ \epsilon_1 \int_{\mathcal{U}_{t*}} |Q_{AB++}|\ \tau_+^3\ \tau_-^{-1}\ d\mu_g.$$

Now, we have

$$Q_{AB++} = \{2\,\gamma\ |\beta|^2 + 2\,\rho\,\alpha - 2\,\sigma\ {}^*\alpha\,\}_{AB}$$

where * denotes duality on $S_{t,u}$. We estimate

$$\begin{aligned}\int_{\mathcal{U}_{t*}} &|(\rho,\,\sigma)\cdot\alpha|\ \tau_+^3\ \tau_-^{-1}\ d\mu_g \\ &\leq \int du \left\{ \tau_-^{-2} \left(\int_{C_u} \tau_-^2\ \tau_+^2\ (\rho^2+\sigma^2) \right)^{\frac{1}{2}} \left(\int_{C_u} \tau_+^4\ |\alpha|^2 \right)^{\frac{1}{2}} \right\} \\ &\leq C\ E \int \tau_-^{-2}\ du \leq C\ E,\end{aligned}$$

comparing with (432). We thus find

$$|\,\text{error integrals 2}\,| \leq C\ \epsilon_1\ E.$$

For the error integrals 3 the worst term is

$$\int_{\mathcal{U}_{t*}} Q_{AB\gamma\delta}\ {}^{(\bar{K})}\hat{\pi}^{AB}\ \bar{K}^{\gamma}\ \bar{K}^{\delta}\ d\mu_g$$

and the leading part of this comes from $(\frac{1}{2}\Phi)^2\tau_+^4 e_+^{\gamma} e_+^{\delta}$, the part of $\bar{K}^{\gamma}\bar{K}^{\delta}$ of the largest weight. Also, ${}^{(\bar{K})}\hat{\pi}_{AB}$ is pointwise bounded by $C\epsilon_1$. Hence, we have

$$\int_{\mathcal{U}_{t*}} |Q_{AB++}\ {}^{(\bar{K})}\hat{\pi}_{AB}|\ \tau_+^4\ d\mu_g \leq C\ \epsilon_1 \int_{\mathcal{U}_{t*}} |Q_{AB++}|\ \tau_+^4\ d\mu_g.$$

Here, the terms $2\gamma|\beta|^2$ and $2(\rho\alpha-\sigma^*\alpha)$ in Q_{AB++} (see formula above) are on equal footing. We estimate

$$\begin{aligned}
&\int_{\mathcal{U}_{t*}} |(\rho,\sigma)\cdot\alpha|\,\tau_+^4\,d\mu_g \\
&\quad \le \int du \left\{ \tau_-^{-2} \left(\int_{C_u} \tau_-^4\,\tau_+^2\,(\rho^2+\sigma^2) \right)^{\frac{1}{2}} \left(\int_{C_u} \tau_+^6\,|\alpha|^2 \right)^{\frac{1}{2}} \right\} \\
&\quad \le C\,E \int \tau_-^{-2}\,du \le C\,E,
\end{aligned}$$

comparing with (433). A similar estimate holds for

$$\int_{\mathcal{U}_{t*}} |\beta|^2\,\tau_+^4\,d\mu_g.$$

We thus find

$$|\,\text{error integrals } 3\,| \le C\,\epsilon_1\,E.$$

Note that in the error integrals 1 the principal part is Q_{AB+-} multiplied by $\tau_+^2\,\hat{\chi}^{AB}$. Thus only the trace-free relative to $S_{t,u}$ part of Q_{AB+-} enters. In any case, the corresponding spacetime trace is 0. That is,

$$\underbrace{\gamma^{AB}\,Q_{AB+-}}_{\text{trace relative to } S_{t,u}} - \underbrace{Q_{+-+-}}_{=4(\rho^2+\sigma^2)} = 0.$$

The absence of an uncontrollable term linear in each of α, $\underline{\alpha}$, $\hat{\chi}$ is an instance of the following general identity: For any three symmetric trace-free 2-dimensional matrices A, B, C we have

$$\operatorname{tr}(A\ B\ C) = 0.$$

Equivalently, there is no product in the space of symmetric trace-free 2-dimensional matrices, because for any two such matrices A, B we have

$$A\ B + B\ A - \operatorname{tr}(A\ B)\ I = 0.$$

The leading role played by symmetric trace-free 2-dimensional matrices can be traced back to the symbol of the Einstein equations. For, as we have seen in Chapter 2, the space of dynamical degrees of freedom of the gravitational field at a point can be identified with the space of such matrices.

Bibliography

[1] R. Arnowitt, S. Deser, C. W. Misner, Energy and the criteria for radiation in general relativity. *Phys. Rev.* **118** (1960), 1100–1104.

[2] R. Bach, Zur Weylschen Relativitätstheorie. *Math. Z.* **9** (1921), 110–135.

[3] R. Bartnik, Existence of maximal surfaces in asymptotically flat space-times. *Comm. Math. Phys.* **94** (1984), 155–175.

[4] R. Bartnik, The mass of an asymptotically flat manifold. *Comm. Pure Appl. Math.* **39** (1986), 661–693.

[5] L. Bel, Introduction d'un tenseur du quatrième ordre. *C.R. Acad. Sci. Paris* **248** (1959), 1297–1300.

[6] H. Bondi, M. G. J. van de Burg, A. W. K. Metzner, Gravitational waves in general relativity. *Proc. Roy. Soc. London Ser. A* **269** (1962), 21–52.

[7] J. Cheeger, D. Ebin, *Comparison theorems in Riemannian geometry*. North-Holland, New York 1975.

[8] Y. Choquet-Bruhat, Théorème d'existence pour certain systèmes d'equations aux dérivées partielles nonlinéaires. *Acta Math.* **88** (1952), 141–225.

[9] Y. Choquet-Bruhat, R. P. Geroch, Global aspects of the Cauchy problem in general relativity. *Comm. Math. Phys.* **14** (1969), 329–335.

[10] D. Christodoulou, Global solutions for nonlinear hyperbolic equations for small data. *Comm. Pure Appl. Math.* **39** (1986), 267–282.

[11] D. Christodoulou, Nonlinear nature of gravitation and gravitational wave experiments. *Phys. Rev. Lett.* **67**, (1991), 1486–1489.

[12] D. Christodoulou, *The action principle and partial differential equations*. Ann. Math. Studies 146, Princeton University Press, Princeton, NJ, 2000.

[13] D. Christodoulou, The stability of Minkowski space-time. In *Surveys in differential geometry: essays on Einstein manifolds*, Surv. Differ. Geom. VI, International Press, Boston, MA, 1999, 365–385.

[14] D. Christodoulou, S. Klainermann, *The global nonlinear stability of the Minkowski space*. Princeton Math. Ser. 41, Princeton University Press, Princeton, NJ, 1993.

[15] R. Courant, D. Hilbert, *Methods of mathematical physics*. Vol. 2: Partial Differential Equations, Wiley Classics Library, John Wiley & Sons, New York, 1989.

[16] L. C. Evans, *Partial differential equations*. Grad. Stud. Math. 19, Amer. Math. Soc., Providence, RI, 1998.

[17] S. W. Hawking, Gravitational radiation in an expanding universe. *J. Math. Phys.* **9** (1968), 598–604.

[18] S. W. Hawking, G. F. R. Ellis, *The large scale structure of space-time*. Cambridge Monogr. Math. Phys. 1, Cambridge University Press, London 1973.

[19] J. Jost, *Riemannian geometry and geometric analysis*. Universitext, second edition, Springer-Verlag, Berlin 1998.

[20] C. W. Misner, K. S. Thorne, J. A. Wheeler, *Gravitation*. W. H. Freeman and Company, San Francisco, CA, 1973.

[21] N. O'Murchadha, J. W. York, Gravitational energy. *Phys. Rev. D* **10** (1974), 2345–2357.

[22] E. Noether, Invariante Variationsprobleme. *Nachr. d. König. Gesellsch. d. Wiss. zu Göttingen, Math-phys. Klasse* (1918), 235–257.

[23] R. Osserman, The isoperimetric inequality. *Bull. Amer. Math. Soc.* **84** (1978), 1182–1238.

[24] R. Penrose, Conformal treatment of infinity. In *Relativité, Groupes et Topologie*, 1963 Summer School Theor. Phys. Univ. Grenoble, Les Houches Lectures, Gordon and Breach, New York 1964, 565–584.

[25] R. Penrose, *Techniques of Differential Topology in Relativity*. Regional Conf. Ser. Appl. Math. 7, SIAM, Philadelphia, PA, 1972.

[26] R. Schoen, S. T. Yau, On the proof of the positive mass conjecture in general relativity. *Comm. Math. Phys.* **65** (1979), 45–76.

[27] R. Schoen, S. T. Yau, *Lectures on Differential Geometry*. Conf. Proc. Lecture Notes Geom. Topology I, International Press, Cambridge, MA, 1994.

[28] R. M. Wald, *General Relativity*. University of Chicago Press, Chicago, IL, 1984.

[29] H. Weyl, *Raum-Zeit-Materie*. Fourth edition, Springer-Verlag, Berlin 1921.

[30] E. Witten, A new proof of the positive energy theorem. *Comm. Math. Phys.* **80** (1981), 381–402.

Index

ADM, 35, 55
angular momentum, 36, 70
  conservation of, 71
Anti-de-Sitter-space, 3
asymptotically
  Euclidean Riemannian manifold, 35
  flat initial data set, 35

Bach tensor, 129
background metric, 48
Bel–Robinson tensor, 105
Bianchi equations, 104
  conformal properties of, 108
  with a current, 126
Bianchi identity, 9, 103
  twice contracted, 9
borderline error integrals, 139
boundary current, 50

canonical equations, 121
canonical form, 122
canonical momentum, 36
canonical stress, 36, 84
Cauchy hypersurface, 6, 86
Cauchy problem, 21
  with initial data on a complete maximal hypersurface, 78
causal curve, 2
causal future, 4
causal past, 4
center of mass integrals, 70, 72
  conservation of, 76
characteristic
  subset, 12
Choquet-Bruhat, 19
Codazzi equations, 32, 78, 113, 131
  trace of the Codazzi equations, 32
  null Codazzi equation, 134
conservation law, 37, 85
conserved quantities, 38
constraint equation, 32, 59, 78
continuity argument, 127
controlling quantity, 125
current, 41
  compatible, 41
curvature tensor, 9, 32

de-Sitter-space, 3
deformation tensors
  estimate of, 127
displacement, 97
  electric, 98
  magnetic, 98
domain of dependence, 24
  theorem, 30

eikonal equation, 110
Einstein, 37
  tensor, 9
Einstein equations, 9, 32, 103
  relative to a maximal time function, 59
Einstein pseudo-tensor, 38
Einstein vacuum equations, 9, 32
Einstein–Weyl energy-momentum conservation laws, 49
electric field, 98
elliptic
  regular ellipticity, 17
energy, 54
  positive energy theorem, 60
  total, 35
energy-momentum tensor, 9, 85
equations of variation, 11
error estimates, 128

Euclidean at infinity, 35
Euler–Lagrange equations, 10, 46, 81, 94, 99
evolution equations, 59
extrinsic geometry, 131

field
  electric, 98
  magnetic, 98

gauge invariance, 39, 74
gauge transformations, 52
Gauss curvature, 133
Gauss equations, 32, 78, 113, 114
  once contracted, 131
  double trace of, 32
Gauss–Bonnet theorem, 113
General Relativity, 9
geodesics
  null, 4
globally hyperbolic, 6
gravitational stress, 81, 84, 85

Hamilton–Jacobi equation, 121
Hamiltonian, 121
Hawking mass, 114, 135
Hilbert's action principle, 37
Hopf–Rinow theorem, 3
hyperbolic
  globally, 6
  regularly, 17, 25
hyperboloid, 92

intrinsic geometry, 131
inverse mean curvature flow, 118
inverted time translations, 110

Killing field, 50

Lagrangian, 10, 36, 42, 79, 94, 122
lapse equation, 59, 131
lapse function, 112, 133
  of the maximal foliation, 131
level set, 110
Lie flow, 41, 44
light cone, 91
linear momentum, 36, 54
Lorentzian metric, 1

magnetic field, 98
mass aspect function, 116, 135, 136
maximal
  development, 25
  spacelike hypersurface, 58
  time function, 59
    canonical, 59
  hypersurface, 'final', 119
Maxwell equations, 13, 99
Minkowski spacetime, 48, 78, 109
  conformal group of, 89
  electromagnetic Lagrangian in, 97
  global stability of, 78
    sketch of the proof of, 103
  non-linear electrodynamics in, 96
motion of matter, 9
  equations of, 9
motion of surfaces
  equation of, 112, 118
  equation of, 136

Noether
  domain Noether current, 42
  Noether current, 37, 46, 84
    conserved, 49
  Noether transform, 17
Noether's theorem, 36, 84
  for sections of $\mathcal{C}$, 47
  in the domain case, 42
null cone, 2
null foliation, 132
null hypersurface, 4, 110
null infinity, 91

optical function, 110, 128

Poincaré group, 49
propagation equation, 134, 136
  for tr θ, 113

regularity, 103
regularly hyperbolic, 17
Ricci tensor, 9

scalar curvature, 9
scale transformations, 110
Schwarzschild metric, 115
Schwarzschild solution, 125
sections of tensor bundles over M, 42
space rotations, 36
space translations, 36
spacelike, 2
spacetime, 1
  curvature, 103
  manifold, 1, 103
spatial rotation group, 110
strongly asymptotically Euclidean, 129
strongly asymptotically flat initial data set, 60, 70, 78
structure equations, 131
symbol, 11

time orientability, 2
time translations, 36, 110
  inverted, 110
timelike, 2

variation equations
  first variation equations, 31
  second variation equations, 31

Weyl, 37
  current, 126
  field, 103, 126
    decomposition of, 137
    electric part of, 126
    magnetic part of, 126
  curvature tensor, 103